Workbook to Accompany Managing Our Natural Resources

6TH EDITION

AGRICULTURAL
FFA
EDUCATION

Workbook to Accompany Managing Our Natural Resources

6TH EDITION

William G. Camp
Betty Heath-Camp

Contributions by **Darold Hehn**

Australia • Brazil • Japan • Korea • Mexico • Singapore • Spain • United Kingdom • United States

***Workbook to Accompany Managing Our Natural Resources*, 6th Edition**
William G. Camp and Betty Heath-Camp

VP, General Manager, Skills and Planning: Dawn Gerrain

Product Director: Matthew Seeley

Senior Director, Development, Skills and Computing: Marah Bellegarde

Product Team Manager: Erin Brennan

Senior Product Development Manager: Larry Main

Associate Product Manager: Nicole Sgueglia

Senior Content Developer: Jennifer Starr

Product Assistant: Jason Koumourdas

Marketing Manager: Scott Chrysler

Senior Director, Production, Skills and Computing: Wendy Troeger

Director, Production, Trades and Health Care: Andrew Crouth

Senior Content Project Manager: Betsy Hough

Senior Art Director: Benjamin Gleeksman

Library of Congress Control Number: 2014946317

ISBN-13: 978-1-305-27315-3

Cengage Learning
200 First Stamford Place, 4th Floor
Stamford, CT 06902
USA

Cengage Learning is a leading provider of customized learning solutions with office locations around the globe, including Singapore, the United Kingdom, Australia, Mexico, Brazil, and Japan. Locate your local office at: **www.cengage.com/global**

Cengage Learning products are represented in Canada by Nelson Education, Ltd.

To learn more about Cengage Learning, visit **www.cengage.com**

Purchase any of our products at your local college store or at our preferred online store **www.cengagebrain.com**

Notice to the Reader
Publisher does not warrant or guarantee any of the products described herein or perform any independent analysis in connection with any of the product information contained herein. Publisher does not assume, and expressly disclaims, any obligation to obtain and include information other than that provided to it by the manufacturer. The reader is expressly warned to consider and adopt all safety precautions that might be indicated by the activities described herein and to avoid all potential hazards. By following the instructions contained herein, the reader willingly assumes all risks in connection with such instructions. The publisher makes no representations or warranties of any kind, including but not limited to, the warranties of fitness for particular purpose or merchantability, nor are any such representations implied with respect to the material set forth herein, and the publisher takes no responsibility with respect to such material. The publisher shall not be liable for any special, consequential, or exemplary damages resulting, in whole or part, from the readers' use of, or reliance upon, this material.

Printed in the United States of America
Print Number: 01 Print Year: 2015

Contents

UNIT V: FISH AND WILDLIFE RESOURCES 123

UNIT VI: OUTDOOR RECREATION RESOURCES 151

UNIT VII: ENERGY, MINERAL, AND METAL RESOURCES 167

UNIT VIII: ADVANCED CONCEPTS 189

PREFACE

This workbook is designed to accompany the sixth edition of *Managing Our Natural Resources* and combines comprehensive practice questions with hands-on activities. Each chapter includes the following features:

Test Your Knowledge

These fill-in-the-blank short answer questions provide practice for learning the key concepts presented in the corresponding chapter in the book.

Activity

These hands-on activities provide information for further exploring, and applying, an important topic presented in the corresponding chapter in the book. They follow a specific format, including a statement of the purpose of the activity, the research that is required, a description of the step-by-step procedural approach, and a set of observations that must be reported at the end of the activity.

In addition, within each section, or set of related chapters, a ***Job Exercise*** is included to help individuals prepare for a career in natural resource management.

ACKNOWLEDGMENTS

Special thanks to Darold Hehn who authored the original edition of the workbook and provided the foundation for subsequent editions.

Photography Credits

Cover images: Sunset on the Beach: © iStockPhoto.com/cmcderm1; Aerial view of water carrying aqueduct in Outer Los Angeles: © iofoto/Shutterstock.com; School of tropical Twinspot snapper, blue background: © Anna segeren/Shutterstock.com; Plowed ground, with brown stones and loose soil: © refleXtions/Shutterstock.com; summer river, © Yuriy Kulik/Shutterstock.com; Glacier Bay National Park on the West Coast of Alaska above the Inside Passage: © JNB Photography/Shutterstock.com; Nordic pine forest in evening light. Short depth-of-field: © Stocksnapper/Shutterstock.com; landscape by the lake in the early morning: © M. Pellinni/Shutterstock.com; Waterfall Kayak Jump, Sangay National Park, Ecuador: © Ammit Jack/Shutterstock.com

Unit Opener images:

Unit I: © Cengage Learning, Source: US Fish and Wildlife Service., Source: US National Park Service

Unit II: Source: Photo by Gary Kramer, USDA Natural Resources Conservation Service., Source: USDA, Agriculture Research Service, photo by Jack Dykinga., Source: Photo by Jeff Vanuga, USDA Natural Resources Conservation Service

Unit III: Source: Photo by Jeff Vanuga, USDA Natural Resources Conservation Service, Source: US Bureau of Reclamation, Dept. of Interior, © Cengage Learning

Unit IV: © Kletr/ Shutterstock.com. Source: USDA, Agricultural Research Service, photo by Stephen Asmus, Source: US Fish and Wildlife Service.

Unit V: Source: USDA Agriculture Research Service, photo by Stephen Asmus., Source: Peter J Carboni/US Fish and Wildlife Service, Source: Donna Dewhurst/ US Fish and Wildlife Service

Unit VI: Source: Bureau of Land Management, Source: Steve Maslowski/US Fish and Wildlife Service, Source: Bureau of Land Management/Wyoming

Unit VII: © huyangshu/Shutterstock.com, © Hand Shiffman/Shutterstock.com, © Wade H. Massie/Shutterstock.com

Unit VIII: Source: US National Park Service, Source: Gary Zahn/US Fish and Wildlife Service, © Alexandra GI/Shutterstock.com

UNIT

I

Introduction

CHAPTER

1

Our Natural Resources

TEST YOUR KNOWLEDGE

Complete the following:

1. In earlier times people ________ to go beyond the boundaries of the land that they knew.
2. The apparently limitless nature of the world led people to believe that our natural resources were endless, boundless, and ________.
3. Now our ________ allow one person to do things that armies of workers could not do in ancient times.
4. A ________ resource is any form of energy that can be used by humans.
5. Natural resources are those things that have not been created by humans that can be used to perform any ________ function.
6. Natural resources are objects, materials (including soil, water, and air), creatures, or energy that are found in nature and that can be used by ________.
7. The ability of an item to be "used by humans" is not ________.
8. The usefulness of many resources changes over time as our science and technology ________.
9. Many things we consider to be ________ today were not resources at all in earlier times.
10. Things were not natural resources because they were not ________ to humans.
11. The United States has a total land area of 3,675,545 square ________.
12. The United States has ________ billion acres.
13. ________ is the uppermost layer of soil, from which we get almost all our food and natural fibers.
14. ________ % of our land is covered by cities, factories, homes, highways, and other artificial structures.
15. Only about ________ % of the total amount of our land is usable for crop production.
16. The soil's major enemy has been ________
17. In the years since our nation began, we have lost ________ of our topsoil to erosion.
18. Another problem we are beginning to face is the ________ of agricultural land to urban or other uses.
19. Once an acre of prime corn-belt land is ________ by concrete or asphalt, it is, to say the least, hard to grow corn there.

20. List the increasingly important uses of our land surface as a natural resource as mentioned in the chapter.
    a.
    b.
    c.
    d.
21. Land-use ________ establishes priorities for land use.
22. ________ % of the earth's surface is covered by water.
23. Water is a natural resource only when it can be put to use by ________.
24. Most of our usable water is always on its way back to the ________.
25. We can increase water's usefulness if we ________ it, for other uses.
26. We must ________ the water.
27. Water was an early source of ________ in this country.
28. Water is needed in even ________ quantities today.
29. Every day Americans use ________ billion gallons of water.
30. Another facet of water resource management is the ________ of excess runoff.
31. There is plenty of water, but most of it is not where people ________ it.
32. List the major water problems that are real and becoming more intense.
    a.
    b.
    c.
33. Fish and wildlife are defined as ________ animals.
34. Fish and wildlife include both game and ________ animals.
35. A ________ natural resource is one that can reproduce itself.
36. Fish and wildlife are considered to be ________.
37. About ________ vertebrate species have become extinct in our nation.
38. About ________ species of animals are regarded as threatened or endangered.
39. Wildlife is not as important for ________ as it is for the pleasure that wild animals, fish, and birds afford us.
40. The structure of wildlife conservation has dictated a program for ________ and sports fishers.
41. A new aspect of fish and wildlife deals with satisfying the demands of the ________ and nonfishing public.
42. When the population of a species starts to fall too low, it may become "________".
43. Threatened species are those that appear to be ________.
44. There are ________ million acres of forest in the United States.
45. ________ of the U.S. forest land is noncommercial and not usable for forest production.
46. U.S. forests have produced ________ billion board feet of timber.
47. A ________ forest canopy is dominated by mature, slow-growing trees.
48. A mature forest produces very little ________ and provides a home for relatively few birds and animals.
49. By good ________, we can cut trees and still have more than before.
50. Most of our energy comes directly or indirectly from the ________.

51. Energy sources include:

    a.

    b.

    c.

    d.

    e.

52. Coal was first discovered in America in ________.
53. The world's known oil reserves have ________ steadily over the years.
54. In 2009, the known crude oil reserves, in the Middle East countries alone, totals over ________ billion barrels.
55. The known crude oil reserves in the United States in 2009 were only ________ billion barrels.
56. A barrel of oil is about ________ gallons or 159 liters.
57. Natural gas is being discovered ________ than it is being used.
58. Widely used metals include:

    a.

    b.

    c.

    d.

    e.

    f.

    g.

59. Undiscovered mineral reserves far ________ what we have already found.
60. As the United States has become richer, our people have found ________ time for recreation.
61. Recreational resources in this country as listed in this chapter include:

    a.

    b.

    c.

    d.

    e.

    f.

    g.

ACTIVITY

Purpose:

List recreational activities in your area.

Research:

1. Obtain information about local outdoor recreation establishments.
2. Identify the outdoor recreation establishments of your area.

Procedure:

List the names of the establishments in their respective areas detailed here:

Forests

Lakes

Beaches

Mountains

Parks

Game Animals

Game Fish

Observations:

1. How many outdoor recreation activities exist in your area?
2. How far away from your home is the nearest recreational activity?
3. What outdoor recreation activity have you participated in recently?

CHAPTER

2

A History of Conservation in the United States

TEST YOUR KNOWLEDGE

Complete the following:

1. The history of this country has been one of ___________.
2. Our ___________ greatness was built up our forests, water, iron, coal, oil, and other natural resources.
3. Our ___________ greatness was built on our soil and water resources.
4. ___________ refers to the careful use of our natural resources to provide as much usefulness as possible to people both now and in the future.
5. Wise management of our natural resources is beginning to replace shortsighted ___________.
6. When the European settlers came, the colonies were covered largely by ___________ forests.
7. The New York Sporting Club was made up of a group of men who hunted primarily for ___________.
8. The New York Sporting Club sought to promote restrictions against ___________ hunters.
9. The first state-administered game and fish commission was created in ___________ in 1865.
10. Native Americans hunted for ___________ and hides.
11. Europeans prized beaver hides and bird feathers, and colonists needed ___________ from game animals and birds.
12. People who killed birds and animals to sell their feathers, furs, and meat became known as ___________ hunters.
13. Passenger pigeons could be killed most easily during ___________ season.
14. Migratory waterfowl could be found most easily on their ___________ grounds.
15. Our ___________ of the land for farming changed the wildlife habitat in extreme ways.
16. The ___________ Act of 1900 made the interstate transportation of game taken against state law a federal crime.
17. In 1918, the Migratory Bird Treaty Act was passed for the protection of ___________ waterfowl.

18. In 1933, Aldo __________ published *Game Management,* a book that forms the basis for what we do today in wildlife management.
19. In __________, the Duck Stamp Act required waterfowl hunters to purchase a $1 stamp.
20. The money generated by the Duck Stamp Act has been used to finance numerous projects to protect and expand North American __________.
21. The United States Fish and Wildlife Service of the Department of __________ was established on June 30, 1940.
22. Today every __________ operates its own fish and wildlife agency.
23. Large and healthy game animal, bird, and fish populations in the United States today are a result of __________ efforts.
24. Probably the earliest recorded timber shortage occurred in __________, about 5,000 years ago.
25. The Romans had to __________ wood from their conquered lands.
26. In 1626, Plymouth Colony passed America's first ordinance controlling the __________ of timber.
27. Several colonies passed laws against __________ of the forest.
28. In the early 1800s U.S. forest preservation efforts centered on saving live oaks for use in building __________.
29. The __________ Forestry Association was organized in 1875 to promote timber culture and forestry.
30. The Division of Forestry was created in __________.
31. Gifford __________ became head of the USDA's Forestry Division in 1898.
32. In 1905, the forest reserves were renamed __________ forests.
33. Pinchot and President __________ greatly expanded the national forest system.
34. During the administration of former President Reagan, large parcels of national forestland were __________ to private landowners.
35. The Weeks Law of 1911 gave the President authority to __________ forestlands for river watershed protection.
36. During the Great Depression, the __________ Conservation Corps (CCC) was involved in forestry work.
37. After World War II, the __________ industry increased America's need for wood.
38. Today we __________ very little of our forest resources.
39. Today we produce __________ wood in this country each year than we use.
40. Our forests are our __________ renewable resource.
41. Removal of forestlands for farm, highway, residential, and industrial uses poses a great __________ to our forest industry.
42. Careful __________ is the key to the future in forestry.
43. Jared Eliot was one of the first to write about soil __________ and drainage in the United States.
44. Since the 1600s, we have lost __________ -third of our topsoil to erosion.
45. Early attempts at soil conservation in this country were __________.
46. In 1935, the Soil Erosion Service became the Soil __________ Service (SCS).
47. Today the SCS is known as the Natural Resources __________ Service (NRCS).
48. Soil __________ districts consist of those interested in the conservation of local soil and water resources.
49. The federal government incentive payments were to assist in the cost of soil conservation and soil- __________ practices.

50. Construction incentive programs, listed in this chapter, that now fall under the USDA's Agricultural Stabilization and Conservation Service (ASCS) are:

    a.

    b.

    c.

    d.

51. When the NRCS helps develop a plan for soil and water conservation, the ASCS helps in ____________ the practices.
52. Soil ____________ is the main side effect of food and fiber production.
53. Our soil is our most important ____________ resource.
54. In the early years, ____________ was the determining factor of where people would live, work, and play.
55. Settlers built homes only where there was adequate ____________.
56. ____________ could be built only where water could be supplied.
57. Water was used as a means of ____________ disposal in many areas.
58. Early interest in water management came out of the ____________ movement.
59. Early emphasis was on water as a channel for transportation and as a ____________ of forestry.
60. The federal government began to assume responsibility for ____________ control.
61. The Flood Control Act of 1936 helped develop and implement plans to reduce ____________ and flooding.
62. Most watershed management projects today are handled through the local soil and water conservation ____________.
63. The ____________ of water used by Americans has grown drastically.
64. Americans have demanded ____________ water.
65. Another problem becoming critical is the ____________ of our water supply and lowered water tables.

ACTIVITY

Purpose:

Evaluate state and federal agencies.

Research:

1. Locate a local telephone book.
2. Find the area detailing state and federal governments.
3. Determine the addresses for the following government agencies, which may be located in your area. The addresses may be found in the telephone book or via the Internet.

Procedure:

Detail the information under the respective headings below.

State Game, Fish, and Parks
U.S. Fish and Wildlife Agency
National Park Service
U.S. Forest Service
State Forestry Agency
U.S. Army Corps of Engineers
Extension Service
U.S. Travel and Tourism
Natural Resources Conservation Service
Agriculture Stabilization and Conservation Service

Observations:

1. What state and federal agencies are located in your town?
2. Which agencies are housed in the same buildings?
3. Which agencies have you visited in the recent past?

CHAPTER

3

Concepts in Natural Resources Management

TEST YOUR KNOWLEDGE

Complete the following:

1. List the components of an environment, as mentioned in this chapter:
   a.
   b.
   c.
   d.
   e.
   f.
   g.
2. We now have the ability to use our resources on unbelievably ________ scales.
3. Many of nature's ________ are not going to last forever.
4. Wise use of our natural resources must become a common ________.
5. ________ our natural resources for the future must become a priority.
6. Everything in our environment could be considered a ________ resource.
7. Natural resources that can last forever regardless of human activities are ________.
8. ________ supplies may be very limited in certain locations.
9. Water supplies may be very ________.
10. Natural resources that can be replaced by human efforts are considered ________.
11. We use more wood and produce ________ wood then we use.
12. Many of our natural resources exist in ________ quantities.

13. Those limited resources that cannot be replaced or reproduced are known as nonrenewable or ________.
14. We can conserve our ________ resource, but once the resource is gone we simply have to do without it.
15. Many exhaustible resources exist in such ________ amounts that they are practically nonexhaustible.
16. There is no practical limit to ________.
17. There is so much ________ ore that there is no practical limit to the metal.
18. There is only so much ________ in the ground.
19. ________ involves the injection of liquid under high pressure into rock layers deep underground to aid in the extraction of oil and gas.
20. We must ________ resources to make them last as long as possible.
21. We can improve existing soil but we cannot ________ soil.
22. Soil is a nonrenewable, or ________, resource.
23. ________ is a science dealing with the complex relationships among living things and their environment.
24. An ________ is any partially self-contained environmental and living system.
25. ________ is a strong concern for the environment.
26. An ________ is a political activist with a special interest in some aspect of the environment.
27. Ecology is based on ________ and objective interpretation of data.
28. An ecologist is a ________.
29. In the role of scientists, ecologists do not attempt to make decisions based on ________ interpretations.
30. Science is ________.
31. When a scientist stops doing science and starts advocating an environmentalist position on political issues, he or she is not talking as a scientist but as an ________.
32. ________ are living subsystems.
33. ________ are nonliving subsystems.
34. Every part of an ecosystem ________ with the other parts of the system and depends on them.
35. All the processes in an ecosystem depend on ________.
36. To have a complete ecosystem there must be three components present:
    a.
    b.
    c.
37. ________ are green plants that produce new food by means of photosynthesis.
38. ________ take the primary source of food, incorporate other chemicals and energy forms, and change it into more complex organic compounds, foods, and tissues.
39. ________ break the organic material back down into their ecosystem.
40. No ecosystem is completely and permanently ________.
41. There is no real "________ of nature."
42. All ecosystems are constantly ________.
43. One species of plant or animal is ________ by another as conditions change and the ecosystem matures.
44. The replacement of one species by another in an ecosystem is ecological ________.
45. If an ecosystem were to become completely stable, the species of plants that would dominate the system would be known as the ________ species.
46. A ________ is the biotic subsystem in an extensive ecosystem.

47. List the major biomes discussed in the chapter.
    a.
    b.
    c.
    d.
    e.
    f.
    g.
    h.
    i.
48. The study of the distribution and residents of the world's biomes is called ________.
49. There is no such thing as a true ________ of nature.
50. ________ in the environment is both continuous and natural.
51. ________ make massive changes in the ecosystem.
52. Managing our natural resources wisely means controlling nature so that we can use its resources without destroying its ________.
53. All living things make up parts of both food ________, and food webs.
54. A food ________ is a sequence of organisms, each of which provides a source of nutrients for the next organism in the chain.
55. A food ________ is a set of overlapping food chains.
56. List examples of food chains from the chapter.
    a.
    b.
    c.
    d.
57. A ________ level can be defined as the number of a given species of plant or animal in a given area at a particular point in time.
58. ________ capacity refers to the ability of an ecosystem to provide food and shelter for a given population level.
59. Population ________ are affected by the availability shelter and predators, as well as diseases and parasites.
60. When a population exceeds its ecosystem's carrying capacity, diseases, predators, or starvation will inevitably ________ the population level.
61. The most profound effect by humans on the ecosystem has been from the advent of ________ production of food and fiber.
62. Farming has drastically increased the carrying capacity of the world for ________.
63. The world human population surpassed ________ billion in 1999.
64. We must begin to better ________ our natural resources.
65. ________ believe in using nature to produce the maximum long-range benefit for people.
66. The National Park Service was established in ________.
67. ________ use encourages us to plan natural resources management activities to produce more than one benefit.
68. The conservation ________ of the United States must be the development and protection of a quality environment that serves both nature and man.

ACTIVITY

Purpose:

Evaluate local natural resources.

Research:

Review and define the following terms:

Renewable

Nonrenewable

Exhaustible

Nonexhaustible

Procedure:

1. Define renewable resources.
2. Define exhaustible resources.

Observations:

1. Place an X inside the box that best explains the resources you researched and defined.
2. Complete this chart by adding local resources and placing an X inside the box that best explains them.

Resource	Nonexhaustible	Exhaustible	Nonrenewable	Renewable
Forest				
Water				
Wildlife				
Gold				
Grass				

JOB EXERCISE FOR CHAPTERS 1–3

Purpose:

Interview a local natural resources employee.

Research:

1. Choose a natural resources position.
2. Research the duties and responsibilities of that position.

Procedure:

1. Write questions to ask in the interview. You may use any of the following:
   a. What are the responsibilities of a person in your position?
   b. What are the rewards of a person in your position?
   c. What are the opportunities for advancement?
   d. What education is required?
   e. What are the employment prospects for this position?
   f. What is the salary range for this position?
2. Interview an employee who works in that position.

Observations:

1. What education do you need in order to be employed in this position?
2. What job responsibility does this position require that you would enjoy doing?
3. What job responsibility does this position require that you would not enjoy doing?

UNIT

II

Soil and Land Resources

CHAPTER

4

Soil

TEST YOUR KNOWLEDGE

Complete the following:

1. Life on this planet is very ___________.
2. List the things we depend on for survival, as mentioned in the chapter.
   a.
   b.
   c.
   d.
   e.
3. No single natural resource is more important to survival than our ___________.
4. The most valuable and productive part of the land is its surface layer, the ___________.
5. ___________ is the layer of natural materials on the earth's surface containing both organic and in-organic material and capable of supporting plant life.
6. Soil is so fragile it can be ___________ almost overnight.
7. Soil is so complex it cannot be ___________ once destroyed.
8. Soil contains four components:
   a.
   b.
   c.
   d.
9. An ideal soil contains about 50% ___________ material and 50% pore space by volume.
10. The ___________ material of soil consists of weathered mineral and rock particles ranging in size from submicroscopic to those readily visible to the human eye.
11. ___________ matter is made up of dead plant and animal materials in varying stages of decay.

12. The __________ of each four main soil components vary from one soil type to another.
13. Soil is formed very __________.
14. __________ materials are materials underlying soil from which the soil was formed.
15. The five categories of soil parent material are:
    a.
    b.
    c.
    d.
    e.
16. __________ are solid, inorganic, chemically uniform, substances.
17. Some common minerals for soil formation are:
    a.
    b.
    c.
    d.
    e.
18. Rocks are __________ of minerals.
19. Rocks are classified into three groups:
    a.
    b.
    c.
20. __________ rocks are formed by the cooling of molten materials.
21. __________ rocks are formed by the solidification of sediments.
22. Almost __________ of the earth's surface is covered by sedimentary rocks.
23. __________ rocks are igneous or sedimentary rocks that have been reformed because of heat or pressure.
24. Much of the Midwestern United States is covered by soils formed from __________ deposits.
25. __________ deposits are generally thought of as wind-blown silt.
26. Alluvial deposits are waterborne __________ left by moving freshwater.
27. __________ deposits were formed on ancient ocean floors.
28. __________ deposits are dead vegetation that gets thick enough to support plant life.
29. __________ soils are made up of recognizable plant materials.
30. __________ soils are more completely decayed so that plant parts are no longer recognizable.
31. __________ is when minerals and rocks are exposed to the weather, and they begin to break into smaller and smaller pieces.
32. __________ and cooling can cause rocks to crack into smaller pieces.
33. Some minerals are __________ and dissolve when exposed to water.
34. As water changes into ice, it __________.
35. Water if it freezes in cracks of rocks can __________ the rock into pieces.
36. __________ may be blown against a large rock by high winds.

37. As __________ move rocks, the rocks grind against each other.
38. As __________ moves soil particles and gravel, the pieces are ground together into smaller pieces.
39. The major weathering forces listed in the chapter are:

    a.

    b.

    c.

    d.

    e.

40. A badly __________ topsoil can be destroyed in a few years.
41. It may take nature thousands of years to __________ the damage through weathering and soil formation.
42. The proportion of organic matter in most soils is __________ to 5%.
43. Organic matter consists of decaying __________ and animal parts.
44. Original tissue is that portion of organic matter that remains __________.
45. Soil __________ is organic matter that is decomposed to where it is unrecognizable.
46. Organic matter serves many important functions. List the functions mentioned in the chapter.

    a.

    b.

    c.

    d.

    e.

47. Most soils have three or more visibly distinct layers or soil __________.
48. Each horizon may have __________.
49. The soil horizons are called:

    a.

    b.

    c.

50. The __________ horizon consists of weathered rocks and minerals.
51. The B horizon contains little __________ matter.
52. The __________ horizon, or topsoil, lies at the surface.
53. The topsoil __________ results from the humus content.
54. Topsoil with more humus is usually __________ than topsoils with less organic matter.
55. As the soil's organic matter content is depleted, its topsoil color becomes __________.
56. The __________ is the most productive part of the soil.
57. List the soil properties.

    a.

    b.

    c.

    d.

    e.

    f.

58. ___________ is the angle of the soil surface from horizontal.

59. Slope is expressed as a ___________ of rise and fall in a given horizontal distance.

60. Texture refers to the proportion of ___________, silt, and clay in the soil.

61. Coarse-textured soils are those with a high proportion of ___________.

62. Fine-textured soils are those with high proportion of ___________.

63. Soil ___________ reflects the natural ability of the soil to allow water to flow through it.

64. Well-drained soils allow ___________ soil water to move fairly quickly out of the plant-growing regions of the soil layer.

65. Poorly drained soil holds excess soil water in the ___________ layers of the soil.

66. Soils with poor drainage will have ___________ or mottled subsoil or even topsoil.

67. ___________ hazard refers to the likelihood that a field will receive flood damage.

68. Even though it may be capable of producing a good crop, a field's long-range potential is ___________ by the probability that the crop will be flooded in any given year.

69. ___________ refers to the degree the soil has already been damaged.

70. A soil that has suffered severe ___________ should be used for agricultural production very carefully.

71. Topsoil and subsoil ___________ refers to the depth of those layers that are available for plant root production.

72. A combination of thin topsoil and thin subsoil will severely ___________ crop production.

73. Land capability classes categorize the ___________ potential of the soil.

74. Class ___________ land is the very best land for agricultural production.

75. The least useful land, from an agricultural standpoint, is Class ___________ land.

76. Very little land is Class ___________, even in the most productive farming areas.

77. Class ___________ land is good land for all types of farming.

78. Class ___________ land can be cultivated and farmed regularly, but it has some important limitations.

79. Class ___________ land may have a strong slope and be subject to severe erosion.

80. Class ___________ land may be quite suitable for pasture, wildlife habitat, or forest production.

81. Class ___________ land can be used for tree production, permanent pasture, or wildlife habitat.

82. Class ___________ land can be used for forest production, wildlife, and recreation.

83. Class ___________ land can be preserved for recreation and wildlife, but it has little agricultural value.

84. The soil ___________ system is used by the Natural Resources Conservation Service (NRCS).

85. Each order of the soil classification system is broken down into ___________, each suborder into great groups, then subgroups, then families.

86. There are about ___________ soil series known in the United States.

87. The characteristics used to determine a soil's series are:

a.

b.

c.

d.

e.

f.

g.

h.

i.

j.

k.

l.

88. The NRCS classification process is known as a soil ___________.

89. The results of a soil survey are known as a soil survey ___________.

90. The soil survey report can be used as a land-use ___________ tool.

ACTIVITY

Purpose:

Describe the soil formation process.

Research:

1. Review information in Chapter 4 of *Managing Our Natural Resources, Sixth Edition* on soil characteristics and formation.
2. Go to the library or search the Internet and find two additional sources.

Procedure:

You are a historian for an environmental magazine. The year is 2456. You are operating a time machine that allows you to see the events of the past. You are viewing an island that protruded above the ocean surface thousands of years ago. As you view the time machine, you see that the island begins as a barren rock, and grows into a beautiful plant- and animal-rich environment.

Observations:

Write a page report on the soil formation process of the island using the following components of the soil formation process.

time	parent material
minerals	rocks
glacial deposits	loess deposits
alluvial deposits	organic deposits
weathering	temperature changes
water action	plant roots
ice expansion	mechanical grinding
organic matter	soil profile
topsoil	subsoil
slope	erosion

CHAPTER

5

Soil Erosion

TEST YOUR KNOWLEDGE

Complete the following:

1. When the first men and women arrived in the area that is now known as America, they found a ________ of abundance.
2. As societies became more complex and the human population grew on this continent, native, wild food soon became inadequate so primitive ________ became necessary.
3. When the Europeans came, they began to ________ more and more of the rich land for farming.
4. Soil erosion is a ________ process.
5. Natural soil erosion is known as ________ erosion.
6. Many minerals and rocks are ________ soluble to some degree.
7. Organic matter is subject to ________ and may be partially water soluble.
8. Removal of the soil's plant covering by ________ has the same effect whether the fire was caused by nature or not.
9. ________ is always a normal part of the soil formation process.
10. ________ erosion is usually much slower than the process of soil formation.
11. As long as soil is ________ by plants, it is protected from excess erosion.
12. The stems and leaves of vegetation ________ down the speed of runoff and wind.
13. The ________ bind the soil particles together.
14. Clearing the natural vegetation away results in a speedup of the normal rate of soil erosion, called ________ erosion.
15. Topsoil that took many thousands of years to ________ can be washed or blown away in just a few years.
16. ________ has made possible all the other things our society has done.
17. Once the land's natural plant cover is removed, the ________ erosion begins.

18. Water causes soil erosion by:

    a.

    b.

19. The first action of water on the soil is that of falling raindrops, or ________ erosion.
20. A raindrop strikes the earth at about ________ miles per hour.
21. If a raindrop hits a plant leaf, its ________ is diminished and breaks into smaller droplets.
22. If a raindrop hits bare earth, that force is absorbed by the soil particles, and then the surface of the soil is ________ apart.
23. When a raindrop strikes, soil particles are ________ in the water and carried away in the runoff.
24. When a raindrop strikes, small particles filter into the soil's surface layer, plugging the soil's ________ spaces, or pores.
25. The second and more destructive action of water on soil is that caused by ________.
26. As surface water begins to move downhill, it picks up ________ particles and carries them along.
27. The faster the water flows the more ________ it becomes.
28. Run-off erosion damage is threefold:

    a.

    b.

    c.

29. Most silt ________ are harmful.
30. List the results of silt deposits:

    a.

    b.

    c.

    d.

    e.

31. Wind has two actions that cause erosion:

    a.

    b.

32. ________ erosion can be a serious problem.
33. As the wind moves along the surface of land, the evaporation of soil moisture is ________.
34. If there are no trees or other surface vegetation to slow the surface wind, the evaporation process is even ________.
35. Finer sand and silt particles, which are very important in a healthy and productive soil, may be ________ away in a cloud of dust.
36. A ________ field is more susceptible than bare soil to drying and blowing.
37. The second action of wind is the effect of ________ soil particles, as smaller soil particles may be carried great distances by the wind.
38. Larger particles of sand and certain clay granules may not be carried away, so the soil surface becomes very sandy or covered by a layer of ________.
39. ________ erosion is the gradual and uniform removal of surface soil.
40. Sheet erosion occurs so ________ that a farmer may not even be aware it is taking place.

41. Sheet erosion is caused by ________ erosion coupled with a slow runoff of soil-laden water.
42. The effect of sheet erosion is to lower the soil's ________.
43. ________ areas in a field probably indicate that sheet erosion is well advanced.
44. ________ erosion is a more rapid and more visible type of erosion.
45. ________ are simply small streamlets cut into the soil surface by running water.
46. As water moves downhill, it ________ away soil particles.
47. Whenever rills appear, erosion has taken place already and the soil has been ________.
48. Left alone, rills will continue to wash out and cause further ________.
49. When a rill becomes very large, it becomes a ________.
50. ________ erosion is less serious than rill and sheet erosion.
51. With just sheet or rill erosion taking place, almost the entire topsoil of a field may be lost before the farmer ________ there is a problem.
52. ________ cause soil removal and sand deposits or drifts.
53. When the soil's vegetative ________ is removed, the surface becomes dry and loose.
54. As the topsoil on a field blows away, the field becomes less ________.
55. As the topsoil on a field blows away, it retains ________ of the rainwater that falls on it.
56. Wind-blown soil ________, or sand drifts, result when the wind is forced to change direction abruptly.
57. The ________ an open field is, the stronger the surface wind is likely to be.
58. The ________ and more flat the field, the more likely it will be affected by wind erosion.
59. Another major problem caused by wind erosion is ________ soil or dust.
60. Airborne soil ranges from the light dust we see on our furniture to great dust ________.
61. When soil particles are removed, the topsoil is ________.
62. When soil particles are ________, if drifts form, further damage is done.
63. No form of erosion is more frightening or awesome than a great ________ storm.
64. No result of erosion is more constant and obvious to everyone than the daily deposit of ________.
65. One of the most severe national disasters in U.S. history occurred in Oklahoma, Arkansas, and the other states in that region, when acres of land were devastated by ________ and blowing dust.

ACTIVITY

Purpose:

Use of the textural triangle.

Research:

The textural triangle shown here can be used to determine the textural class of a soil sample. In order to use the triangle, you must first know the percentage of sand, silt, and clay in your sample. The percentage of sand, silt, and clay can be calculated by mechanical analysis.

Once you know the percentages, locate the percent sand along the bottom of the triangle, the percent clay along the left side of the triangle, and the percent silt along the right side of the triangle. The textural class is determined at the point where the clay, silt, and sand intersect. If all lines intersect on a division line between the classes, move toward the finer textured soils.

Soil Textures and Their Particles

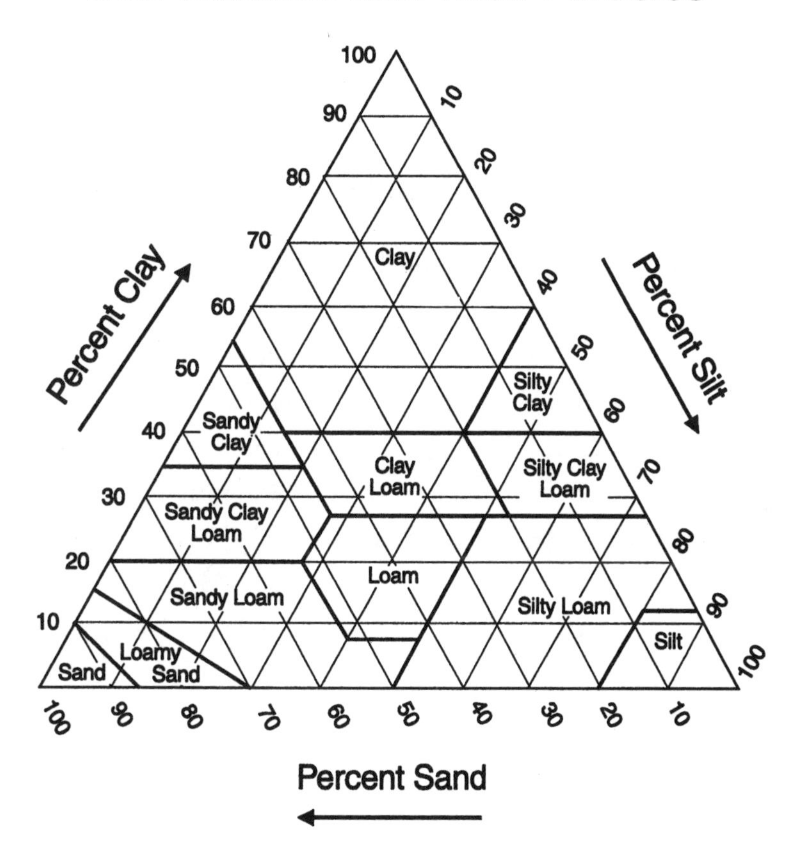

Procedure:

Given the relative amounts of sand, silt, and clay, find the textural classes of the following:

a. 25% sand ________
   20% clay
   55% silt

b. 50% sand ________
   40% clay
   10% silt

c. 70% sand ________
   25% clay
   5% silt

d. 40% sand ________
   42% clay
   18% silt

e. 10% sand ________
   80% clay
   10% silt

Observations:

1. What percent range of silt must a soil contain to be called a
   a. Silt clay loam ________
   b. Silt loam ________
   c. Silty clay ________
   d. Silt ________
2. What percent range of sand must a soil contain to be called a
   a. Sandy clay ________
   b. Sandy clay loam ________
   c. Sand ________
   d. Loamy sand ________
3. What percent range of clay must a soil contain to be called a
   a. Clay loam ________
   b. Clay ________
   c. Sandy clay ________
   d. Sandy clay loam ________

CHAPTER

6

Controlling Erosion on the Farm

TEST YOUR KNOWLEDGE

Complete the following:

1. ________ or geological erosion takes place all the time.
2. Soil erosion cannot be completely stopped or ________.
3. There are ________ land capability classes.
4. Class ________ land is the very best for agricultural production.
5. Class I land is ________, fertile, and relatively safe from erosion.
6. Class ________ land is very rough, rocky, covered with deep sand, or under water.
7. Knowing the capability class of a piece of land may not be necessary, but using land ________ is crucial.
8. The use of land within its ________ is the most important step in land use.
9. Class ________ land can be used safely for any kind of farm production.
10. Class I land is ________ generally subject to erosion or flooding.
11. Class I land can be ________ every year with little damage.
12. Class ________ land has a gentle slope and is subject to some erosion.
13. ________ of Class II land can be done with little danger of excessive soil erosion.
14. List the erosion-control techniques for Class II land that are mentioned in the chapter.
    a.
    b.
    c.
    d.
    e.

15. Class III land has a moderate slope and is subject to severe ________.
16. If Class III land is used for occasional cultivation, what erosion-control measures are required?
    a.
    b.
    c.
    d.
    e.
17. If Class III land is not sloping, it may have serious ________ problems or droughtiness.
18. Class ________ land is too steep, sandy, rocky, or wet for safe and regular cultivation.
19. Class IV land should be covered by ________ almost continuously.
20. No-till planting is best for Class IV land but if cultivation is done, what measures will be needed?
    a.
    b.
    c.
    d.
    e.
    f.
21. Class V land may be too ________, too sandy, or too rocky for cultivation.
22. Class V land is level or nearly level, but its other limitations make it suitable only for permanent ________ or forest production.
23. Class VI land has what very limiting factors that make it unsuitable for cultivation?
    a.
    b.
    c.
24. Class VI land can be safely used for permanent pasture once a good ________ cover is established.
25. ________ should be avoided on Class VI land.
26. ________ production is a good use for Class VI land.
27. Class VII land is so ________ that its use for grazing or even forest production may be limited.
28. Class VIII land is so rough that any existing vegetation should be ________.
29. Water causes soil erosion mostly through what two actions?
    a.
    b.
30. Erosion-control techniques seek to ________ the damage caused by the two soil erosion actions.
31. There are two categories of erosion control:
    a.
    b.
32. Erosion is best controlled when vegetative and mechanical measures are used ________.
33. During the growing season, cultivated land is afforded some protection from water erosion by the ________ crop.

34. List the forms of vegetative erosion control:
    a.
    b.
    c.
    d.
35. On fields where erosion may be severe, close-growing ________ crops may provide protection from erosion.
36. Cover crops may improve the soil's ________ matter content as well as its structure and tilth (physical condition).
37. If the cover crop is plowed under at the beginning of the next regular production cycle, it is called green ________ crop.
38. Close-growing ________ and clovers make good cover crops.
39. A crop ________ is an orderly and repeated sequence of different crops grown on the same field.
40. A common 3-year crop rotation in parts of the Midwest might be ________.
41. A 4-year rotation might be corn-cotton-oats-hay (or ________).
42. The chapter discusses advantages of crop rotation over continuous row cropping. List those advantages.
    a.
    b.
    c.
    d.
    e.
    f.
    g.
43. ________ cropping is the production of alternating bands of different crops.
44. On sloping land, the strip cropping strips are laid out either:
    a.
    b.
45. A typical ________ scheme would alternate row crop, hay, row crop, hay.
46. The effect of strip-cropping is to provide bands of ________ vegetation alternated with row-cropped bands.
47. List the values of heavier vegetation, as mentioned in the chapter.
    a.
    b.
    c.
    d.
    e.
48. Because rainfall or snowmelt produces more surface water than the soil can absorb at one time, ________ occurs.
49. The collecting water must be led safely out of the field, or its uncontrolled movement could lead to the very rapid formation of ________.
50. The best place to put grassed waterways is in ________ drainage ways.
51. A grassed ________ is a drainage way permanently covered by vegetation.
52. In a terraced or contour row-cropped field, the rows tend to conduct collecting runoff in some specific direction and is ________ into a grassed waterway.

53. The most common type of grassed waterways is the ________ -shaped waterway.
54. Once the waterway has been formed, it must be quickly protected by a close-growing ________.
55. The seedbed of a grassed waterway should be ________ seeded, from two to three times the normal rate.
56. The objective of a grassed waterway is to get a ________, lush growth of grass to produce a firm sod.
57. Once the sod is produced, the grass of a grassed waterway may be used for ________.
58. Once the sod of a grassed waterway is established, the waterway is not ________ along with the rest of the field.
59. If the sod of a grassed waterway is damaged, it should be ________ immediately.
60. The earliest tillage was the use of a hand-held ________ to gouge the earth so that a seed could be planted.
61. List the purposes of tillage:

    a.

    b.

62. As animal power and tractor power was developed, tillage came to mean the complete ________ of the vegetation covering the soil as a step in seedbed preparation.
63. ________ tillage is any tillage system that is economically practical for crop production and aids in soil and water conservation.
64. Conservation tillage replaces the older methods of ________ tillage that left bare soil exposed during the nongrowing season.
65. Conservation tillage depends on the crop ________ from the previous crop to protect the soil from erosion until the new crop is well established.
66. Common conservation tillage techniques are:

    a.

    b.

    c.

    d.

    e.

    f.

    g.

67. ________ farming means plowing "around" the hill instead of up and down it.
68. Contour farming conducts the ________ across the slope to a grassed waterway or tree line where it can be safely controlled.
69. ________ are large surface channels constructed on the contour with a controlled rate of fall.
70. Terraces are designed to accept the ________ and conduct it across the slope to some protected area.
71. Terraces are needed on fields whenever the slope exceeds about ___% and where the slope is over a few hundred feet long.
72. A ________ is a row of trees and/or shrubs planted across the prevailing wind direction.
73. Windbreaks are normally planted to ________ the farm buildings and residences from high winds.
74. ________ are planted to provide protection for crops and livestock.
75. With shelterbelts, some protection is offered downwind at least ________ times the height of the shelterbelt.
76. Crop ________ left on the field helps conserve soil moisture, reduces surface wind speed, and holds loose soil particles in place.
77. It is possible to row crop on the contour for ________ much like for water.

ACTIVITY

Purpose:

Evaluate water/wind erosion control methods.

Research:

1. Review the erosion control methods for water and wind erosion.
2. Determine whether the controls are vegetative or mechanical.

Procedure:

Define the following erosion control methods:

1. crop rotation
2. strip cropping
3. conservation tillage
4. cover crops
5. grassed waterways
6. contour farming
7. terrace
8. shelterbelt
9. windbreaks

Observations:

1. Which of the conservation methods are vegetative control?
2. Which of the conservation methods are mechanical control?

CHAPTER

7

Nonfarm Erosion Control

TEST YOUR KNOWLEDGE

Complete the following:

1. Soil erosion can become a major problem whenever the soil is ____________.
2. Daily, as many as ____________ acres of U.S. farmland are converted daily to nonfarm use.
3. Millions of acres of land are disturbed annually by strip mining for ____________ and other resources.
4. As new highways are constructed and old ones are widened, moved, or repaired, millions of miles of road banks are subject to ____________ erosion.
5. ____________ involves more extreme changes on the soil surface than farming does.
6. Controlling nonfarm erosion involves:

   a.

   b.

   c.
7. The two types of nonfarm erosion control practice discussed in this chapter are:

   a.

   b.
8. Because of the ____________ nature of most construction, more expensive erosion control practices are often economically essential.
9. A ____________ ditch or berm of earth placed across the slope collects runoff and conducts it to a site for safe disposal.
10. The diversion ditch or berm may be sodded or covered by ____________.
11. ____________ collect excess soil water in the form of runoff and conduct it elsewhere for disposal.
12. It is not uncommon to find ____________ waterways in built-up areas.

13. Concrete is used for construction of waterways because:

 a.

 b.

 c.

14. Where the slope is very steep, a concrete-lined waterway is called a water ____________.

15. In urban and commercial development, ____________ deposits must be removed with costly-machinery and labor.

16. A ____________ basin is designed to be filled up by sediment or mud.

17. The muddy water of the sediment basin is held long enough for the sediment to ____________ out in the form of mud.

18. Sediment basins are normally ____________ structures.

19. Where banks will be particularly steep, ____________ or stone walls may be the answer.

20. Longer or less steep banks are protected by:

 a.

 b.

 c.

 d.

 e.

21. ____________ may start as a slow trickle at the top of a long, steep road bank and become a racing stream before it reaches the bottom.

22. Each terrace breaks the long ____________ into shorter slopes.

23. Vegetative erosion controls mentioned in this chapter include:

 a.

 b.

 c.

 d.

 e.

24. The most common vegetative control technique for erosion control on nonfarm sites is the ____________.

25. Lawns are vegetative control that are established by:

 a.

 b.

 c.

 d.

26. The seedbed should be ____________ by thorough tilling, smoothing, and removal of rocks and other debris.

27. ____________ is a living layer of mature grass and topsoil that usually is found in squares or long, rolled strips.

28. Sod is an almost ____________ lawn.

29. Sod is much more ____________ than seeding, or plugging a lawn to grass.

30. When grass is to be established by seeding, light ____________ may be necessary.

31. List the things mentioned in this chapter that mulch does:

    a.

    b.

    c.

32. ____________ are used as chemical bonding agents, which serve the same function as mulch.
33. Mulches include:

    a.

    b.

    c.

    d.

34. Low-growing ____________, vines, or other plants can be used much like grass is used in a lawn.
35. Cover ____________ on construction sites are used to protect an area temporarily.
36. If construction is ____________ for any reason, a cover crop may be in order.
37. Winter ____________ might provide a temporary cover crop.
38. Virtually all the rain falling on a ____________ either evaporates or runs off.
39. Soil washed from the banks may cover parts of the highway with ____________.
40. Erosion in the ____________ underneath the highway could cause sags and cracks in the road.
41. Erosion-control techniques along highways are basically the same as those used elsewhere; however, the ____________ is much greater.
42. During the last quarter of the last century, a large increase occurred in the strip mining of ____________.
43. Because of the high impact on the environment, numerous federal and state ____________ control mining.
44. Strip mining is regulated in every ____________ in which it occurs.
45. List the differences of each strip mine, as discussed in this chapter.

    a.

    b.

    c.

46. By law, each mining operation must have a ____________ plan.
47. The surface above the coal, the ____________, must be placed in a stabilized state.
48. After the coal is extracted, there is a 2-year period within which the land must be ____________.

ACTIVITY

Purpose:

Calculate the topsoil needed.

Research:

The most common vegetative control technique for erosion control on nonfarm sites is the lawn. Sometimes topsoil must be applied before a lawn can be established. Topsoil is purchased by the cubic yard.

To determine the yards of topsoil needed, use the following formula:

$$\frac{\text{Length} \times \text{Width} \times (\text{Depth in inches})}{12} = \text{Cubic feet}$$

$$\frac{\text{Cubic feet}}{27} = \text{Cubic yards}$$

Example:

$$\frac{20 \times 10 \times 3}{12} = 50 \text{ Cubic feet}$$

$$\frac{50}{27} = \text{Cubic yards}$$

Procedure:

Calculate the topsoil needed for the following lawn areas:

1. A residential lawn measuring 100 feet long and 50 feet wide. The owner wants to add 4 inches of topsoil.
2. A commercial landscape with the following dimensions:
   a. Area #1: 50 feet long and 25 feet wide, 4 inches deep
   b. Area #2: 29 feet long and 8 feet wide, 2 inches deep

Observations:

1. What would be the cost for procedure Area #1 if topsoil cost $10.00 per yard?
2. What would be the cost for procedure Area #2 if topsoil cost $12.00 per yard?

CHAPTER

8

Rangeland Management

TEST YOUR KNOWLEDGE

Complete the following:

1. The ____________ and importance of rangelands on a worldwide basis today can hardly be overstated.
2. ____________ refers to the land or water area that provides food and habitat for animals (including birds and fish).
3. ____________ or range, grassland, or prairie refers to the land areas of the world that tend to be naturally covered by grasses, grass-like plants, forbs, and shrubs as the primary vegetation instead of trees.
4. Natural rangelands, as mentioned in the chapter, occur in what form?

   a.

   b.

   c.

   d.

   e.

   f.

   g.
5. Human-generated rangelands consist of ____________ land and forage crop land produced on areas that were originally covered by forest or natural grasslands.
6. Of the total land area on earth, approximately ____________ % is managed as permanent rangeland.
7. West of a line drawn between Texas and Illinois is a general area frequently referred to as the Great ____________.
8. Rangeland makes up about ____________ of all the land in the United States.
9. Rangeland in the United States is about ____________ million acres.
10. The most ____________ and accessible rangelands are generally privately owned.

11. The federal rangelands are managed by what agencies?

 a.

 b.

 c.

12. The Bureau of Land Management manages most of the federal land, ____________ it to ranchers for grazing livestock.
13. Grasslands consist of three broad types:

 a.

 b.

 c.

14. Because settlers recognized the ____________ of the land and cleared it for farm production, the tallgrass prairies are almost gone today.
15. All of the shortgrass region requires ____________ to make cultivated farming profitable.
16. Almost all of the original tallgrass prairie land has been ____________ to farm production or other human developments.
17. The African ____________ is a vast and rich grassland.
18. In parts of South America, there are great regions of ____________.
19. In Brazil, the grasslands are called ____________.
20. In other parts of the Americas, the term used to describe some grasslands is ____________.
21. In Spain, grassland is the ____________.
22. In the Philippines, grassland is the ____________.
23. Before the arrival of the European explorers and settlers, the center of this continent was a land of vast, uninterrupted ____________.
24. Whenever ____________ had destroyed the forest, rapidly growing grasses would quickly replace the trees until new trees returned.
25. Often fires were set by ____________ Americans for hunting.
26. When the first European settlers arrived, the only checks on the population of herbivores were:

 a.

 b.

 c.

27. To harvest enough wild game, Native Americans often resorted to the use of ____________ fires.
28. A "____________ party" would be set up by Native Americans.
29. Range fires killed ____________ that might otherwise have become dominant.
30. A rangeland without shrubs and trees to compete with grasses for sunlight, nutrients, and water produces lush grazing and has a much higher ____________ capacity for herbivores.
31. Beyond the effect of the human-caused and natural fires, on the carrying capacity of the grasslands, were

 a.

 b.

32. ____________, in the form of periodic drought, coupled with years of unusually heavy rainfall produced, reduced grazing during some years and abundant grazing during other years.
33. Early pioneers spoke of herds of "____________" that stretched as far as the eye could see.

34. ____________ explorers released donkey, cattle, and other grazing animals into the wild.
35. Spanish animals found a land of plentiful and nutritious grazing and the result was ____________.
36. The earliest ____________ came to the grassland to look for gold, silver, or to harvest fur pelts.
37. Most of the new settlers came with the specific intent of earning a living from ____________.
38. Tall, abundant grass offered free grazing for millions of ____________.
39. Cattle ranchers had little ____________ of the impact of their activities on the ecosystem.
40. The motive of the cattle ranchers was ____________, not conservation.
41. Cattle populations were pushed up beyond the ecological ____________ of the ranges.
42. The winter of ____________ −1886 was one when cattle died or became sick as a result of unusual cold combined with the lack of adequate grazing.
43. ____________ wire was invented in the 1870s.
44. Barbed wire was violently ____________ by the cattle ranchers.
45. ____________ were seen by the cattle ranchers as invasions of "their" land.
46. Cattle ranchers came to understand that range restoration was not possible without fencing to -control ____________.
47. List the reasons from the chapter, why overgrazing continued even when the range was settled and fenced:

    a.

    b.

    c.
48. The Taylor Grazing Act of 1934 was designed to prevent continued ____________ in the arid grasslands.
49. Responsibility for the administration of the Taylor Grazing Act was passed to the Bureau of ____________ Management (BLM) in 1946.
50. The concept of the grazing ____________ has become an effective tool for management of government rangelands.
51. As of 2010, the BLM controlled a total of ____________ million acres of public lands.
52. Only ranchers who own land adjacent to grazing districts are eligible to ____________ land from BLM-managed ranges.
53. Ranchers who hold leases to BLM-grazing rights can graze their livestock at very ____________ costs on public domain grasslands.
54. Ranchers who hold leases must abide by strict management policies set by BLM officials based on the recommendations of government range ____________.
55. Major types of vegetation in the rangelands include:

    a.

    b.

    c.

    d.
56. Grasses are valuable from two standpoints:

    a.

    b.
57. The ____________ of grasses on rangeland varies substantially from year to year.
58. Grazing ____________ affects the relative amounts of the four major types of vegetation present.

59. ____________ grazing benefits the grasses in comparison to other types of plants.
60. As long as less than ____________ of the length of a grass stalk is eaten, grazing causes little damage to the grass plant.
61. ____________ grazing is detrimental to the grasses.
62. When most of the grass ____________ is eaten, it is more difficult for the plant to recover.
63. Varieties of grass that are easily damaged by moderate grazing are called ____________.
64. Decreasers are often the grasses that grazing animals find most ____________.
65. The more desirable grasses tend to be eaten until they are ____________.
66. Rangeland plants that tend to thrive under heavy grazing are called ____________.
67. Many increasers are successful merely because the grazers find them ____________ and so avoid eating them.
68. Plants that move into an area after it has been badly overgrazed are called ____________.
69. ____________ areas are also subject to being taken over by less desirable plants.
70. The ideal situation involves a ____________ level of grazing.
71. A rangeland without grazing as well as with overgrazing will produce less total ____________ than one with moderate grazing.
72. The primary objective of range management is the long-term ____________ of livestock productivity from managed rangeland.
73. List additional objectives for range management:

    a.

    b.

    c.

    d.

    e.

    f.

74. The first step in a planned range management program is the determination of the ____________ capacity or grazing capacity of the area.
75. From the chapter, list the factors that affect carrying capacity:

    a.

    b.

    c.

    d.

    e.

    f.

    g.

76. Grazing capacity is used to determine an acceptable ____________ rate.
77. An area's stocking rate is expressed in terms of animal ____________ units (AEU).
78. One AEU is the amount of forage that is required to feed a ____________ -pound animal for a given period of time.
79. One ____________ for a month is known as an animal unit month.

80. Complete the following.

    a. 1 steer = ____________ AEU

    b. 5 sheep/goats = ____________ AEU

    c. 1 horse/bull = ____________ AEU

    d. 1 elk = ____________ AEU

    e. 4 deer = ____________ AEU

81. Once the grazing capacity of an area is determined, grazing rates must ____________ to ensure that the rate is not exceeded.

82. Undergrazing and overgrazing discourage the growth of ____________ grasses.

83. Common management systems include:

    a.

    b.

    c.

84. In ____________ grazing the livestock are kept in the same area year-round and are allowed to graze as they choose.

85. In ____________ grazing a range area is fenced into two or more separate grazing areas.

86. ____________ management is when animals are moved every few days to a new paddock according to a predetermined schedule or as the grass appears to reach the point where it needs to be allowed to recover.

87. ____________ grazing is the first step in range restoration.

88. Grasses ____________ to the area have been found to be the only long-term solution for reseeding.

ACTIVITY

Purpose:

Categorize range plants.

Research:

1. Locate your state's Range and Pasture Judging Handbook or similar materials.
2. Locate the list of plants from your instructor.
3. Review the information about the following range plant characteristics:
   a. Grasses
   b. Sedges (grass-like)
   c. Forbs
   d. Shrubs
4. Review the following response characteristics:
   a. Desirable
   b. Undesirable
   c. Invader

Procedure:

1. Define "increaser."
2. Define "decreaser."
3. Define "invader."

Observations:

Complete this table by placing a check mark in the boxes that best explain the plant.

Plant name	Tree	Grasses	Sedges	Forbs	Shrubs	Native	Introduced

CHAPTER

9

Landfills and Solid Waste Management

TEST YOUR KNOWLEDGE

Complete the following:

1. One undeniable problem facing our society is that of solid ____________ disposal.
2. ______________ waste management may be our most pressing remaining environmental problem.
3. Solid waste is ______________, nonsoluble materials ranging from municipal garbage to industrial wastes.
4. List the two major sources of solid waste from the chapter.

   a.

   b.
5. ______________ solid waste consists of spoilage from mining, logging, and other industrial processes, not disposed of in landfills.
6. ______________ is the largest producer of industrial solid waste.
7. ______________ mining is the process of excavating the coal seam by removing the soil and rocks above it and then digging out the coal.
8. At one time ______________ was considered a solid waste, but now it is a by-product that many home and business owners prize as bark mulch.
9. The total amount of municipal solid waste generated has almost ______________ between 1960 and 2011.
10. A typical fast food restaurant produces about ______________ pounds of solid waste per day.
11. The United States produces the ______________ total amount of municipal solid waste (MSW), as well as the largest amount of MSW per capita basis.
12. The three general types of solid waste are:

    a.

    b.

    c.

13. Hazardous waste is a waste with properties that make it dangerous or potentially ______________ to human health or the environment.
14. Wastes are considered hazardous if they are:

    a.

    b.

    c.

    d.
15. ______________ wastes are subject to spontaneous combustion or have flash points below 140 degrees Fahrenheit.
16. ______________ waste has acids with pH of 2.0 or lower or bases of pH 12.5 or higher.
17. ______________ waste is subject to explosion or produces toxic fumes or gasses under heating or when compressed or exposed to water.
18. ______________ waste is harmful or fatal when swallowed or absorbed into the body through contact with the skin or eyes.
19. Radioactive waste is of what three types?

    a.

    b.

    c.
20. ______________ waste (LLW) includes medical waste from hospitals.
21. ______________ waste (HLW) is made up of "irradiated" or used nuclear reactor fuel.
22. ______________ mill tailings are the residues that are left over after uranium ore extract is processed to produce uranium and thorium.
23. ______________ waste is municipal waste and industrial waste that does not meet the definition of radioactive waste.
24. Human societies have managed their solid waste for centuries by ______________ it.
25. Studies revealed that significant amounts of water ______________ through landfills.
26. Water that enters a landfill and moves through the buried material is known as ______________.
27. The leachate percolates downward, until it either flows out of the landfill as seepage or enters the ______________.
28. Once contaminants are in the groundwater, they remain there ______________ or until the water is removed for human use.
29. Leachate from buried solid waste is one important reason that well water must be tested periodically for ______________.
30. A ______________ is an open area into which garbage is placed.
31. Landfills are covered by a layer of material, typically ______________.
32. The two basic types of landfills are:

    a.

    b.
33. The natural attenuation landfill is designed to allow ______________ percolation of precipitation to pass through the waste and the underlying soil.
34. The expectation in a natural attenuation landfill is that the leachate will be attenuated (neutralized) by the ______________ and soil particles.

35. Municipal solid waste can be of three kinds:

    a.

    b.

    c.

36. Paper products and yard waste are nonhazardous and can be ______________ disposed of in a natural attenuation landfill.

37. List the three basic landfill shapes:

    a.

    b.

    c.

38. An ______________ landfill is used when waste is placed at or nearly at the normal soil level.

39. A ______________ fill landfill involves pushing the waste over the side of a hill or slope, then pushing a covering layer over it.

40. The most common kind of landfill is a ______________ fill landfill.

41. A ______________ fill landfill is one in which a trench is formed by removing the soil, the waste is placed in the trench, and the soil that had been removed is used to form the final cover.

42. List the six natural mechanisms that occur in the soil as leachate percolates through.

    a.

    b.

    c.

    d.

    e.

    f.

43. ______________ is the process by which the molecules of a chemical adhere to the surface of some other material.

44. A given volume or weight of ______________ has a huge surface area compared to the same volume or weight of sand or gravel.

45. Contaminants in leachate "stick to" ______________ particles so tightly that they are permanently -removed from the percolating water.

46. ______________ removal takes place when bacteria, fungi, or other soil microorganisms break down or absorb the leachate constituents.

47. Soil ______________ use for food some of the materials we would consider to be contaminants.

48. ______________ exchange neutralizes the leachate by changing the molecular structure of the ions.

49. ______________ means that the leachate is decreased by mixing it with large quantities of water.

50. ______________ involves the physical removal of solid constituents from the leachate by trapping them in pores in the soil.

51. ______________ is the process of a phase change in the leachate, wherein the liquid contaminant in the leachate becomes a solid material that is removed.

52. A ______________ landfill is designed to minimize the seepage of leachate into the surrounding soil and groundwater.

53. Total containment is necessary for certain hazardous and ______________ waste.

54. ______________ materials are disposed of in deep shafts, salt beds, and other extreme locations.

55. A natural attenuation landfill can be made into a containment landfill by the addition of some sort of ________________.
56. The liner can be natural, compacted ________________ or bentonite soil or by using a synthetic material.
57. The way to minimize seepage is by using ________________ layers.
58. The more ________________, the less seepage will occur over time.
59. Leachate collection ________________ can be used to capture the seepage collecting above the liners and drain it away.
60. The leachate is ________________ in a wastewater treatment plant.
61. Containment landfills are very ________________ to build.
62. Leachate must be captured in the drainage pipes and treated for ________________ to 50 years before the landfill is safe to simply abandon.
63. The primary reason recycling is important is the ________________ of solid waste disposal.

ACTIVITY

Purpose:

Identify landfill characteristics.

Research:

1. Locate Internet Web sites that detail information about landfills. (http://www.howstuffworks.com/environmental/green-science/landfill.htm)
2. Answer the following questions relating to landfills.

Procedure:

1. What is a landfill?
2. What are the elements of a landfill?
3. What is a liner?
4. What are the different types of liners?
5. What is a cover?
6. What are some problems with covers?

Observations:

1. List the Internet Web site sources you used to answer the questions.
2. Detail two "search" words you used to locate landfill information on the Internet.

CHAPTER

10

Wetland Preservation and Management

TEST YOUR KNOWLEDGE

Complete the following:

1. Wetlands are a premier, underrated, and overlooked natural ____________________.
2. What is the ecological role of wetlands?
   a.
   b.
   c.
   d.
   e.
3. One-third of all threatened and endangered species live only in ____________________.
4. Wetlands exhibit what three characteristics?
   a.
   b.
   c.
5. An area whose hydrology includes frequent saturation with free water is probably a ____________________.
6. Plant types that are attracted to water-saturated growing conditions are called ____________________.
7. ____________________ soils are soils that tend to be saturated with water most of the time.
8. Hydric soils may be mottled with white or ____________________ or may be very yellow.
9. An area with hydric soils is probably a ____________________.

10. The U.S. Army Corps of Engineers defines wetland as an area:

    a.

    b.

    c.

11. The Corps is responsible for __________________ wetlands.
12. Wetlands as defined by the U.S. Fish and Wildlife Service are lands transitional between terrestrial and aquatic systems where the water __________________ is usually at or near the surface or the land is covered by shallow water.
13. The U.S. Fish and Wildlife Service wetlands must have one or more of the following three attributes:

    a.

    b.

    c.

14. __________________ wetlands may or may not have all the three characteristics of the jurisdictional wetland definition.
15. Since the beginning, wetlands were considered a __________________ rather than an asset.
16. The U.S. government authorized and subsidized __________________ of wetlands beginning in 1849 with the passage of the Swamp Lands Act.
17. For over a century, the process of draining wetlands was thought of as the __________________ of swampland for productive use.
18. In 1972, the federal government changed its policy about wetlands and their importance and took steps to protect and restore wetlands by requiring __________________.
19. Under the Wetland Conservation provision, farmers are required to __________________ the wetlands on their farms and ranches in order to be eligible for the U.S. Department of Agriculture (USDA) farm program benefits.
20. List the two most common techniques for wetland identification:

    a.

    b.

21. Off-site inspection involves checking maps and wetland __________________ maintained by the relevant federal agencies.
22. Detail the three principal wetland resources available to the landowner, as described in the chapter.

    a.

    b.

    c.

23. The NWI is concerned with wetlands and __________________ water in the United States.
24. The Natural Resources Conservation Service (NRCS) maintains a list of __________________ soils in the country with soil survey maps.
25. The U.S. Geological Topography Maps look at __________________ covers, surface characteristics, and bogs and marshes.
26. If in an on-site identification the site in question concerns __________________, filling, or discharge into a suspected wetland, the U.S. Army Corps of Engineers will be called upon.
27. If in an on-site identification the site is close to a __________________ or inland water, the U.S. Fish and Wildlife Service will be called.
28. The NRCS from the USDA is needed to identify wetlands on agricultural lands or nonagricultural lands that border __________________ lands.

29. By looking at the plants present, the condition and development of the soil, and the water makeup of the area, a technician can determine if the site should be classified as a ___________________ area.
30. List the types of wetlands.
    a.
    b.
    c.
    d.
    e.
    f.
    g.
    h.
31. List the types of marshes.
    a.
    b.
    c.
32. List examples of marsh plants.
    a.
    b.
    c.
    d.
    e.
33. ___________________ are those areas that border rivers, lakes, and streams and that are flooded periodically.
34. List examples of wildlife found in ponds.
    a.
    b.
    c.
    d.
    e.
    f.
    g.
35. List the types of swamps.
    a.
    b.
    c.
    d.
36. ___________________ are areas that are very damp, usually with evergreens present, with a floor covered with moss and peat.
37. Prairie ___________________ are usually full in the spring and early summer before water levels start to drop off.
38. ___________________ pools last for only a few months each year.
39. In the 1600s, there were over ___________________ million acres of wetlands in the United States.

40. Detail the major causes of the loss of wetlands, as mentioned in the chapter.

    a.

    b.

    c.

    d.

    e.

41. Because wetlands were considered low-value land for so long, roads and bridges were constructed through wetlands rather than ___________________ them.

42. ___________________ has been a major factor in the loss of wetland areas.

43. The most common mining practice that affects wetlands is the mining of ___________________ moss.

44. The lower 48 states have lost over ___________________ of their original wetlands.

45. Wetland preservation efforts can be classified into three major areas.

    a.

    b.

    c.

46. ___________________ wetlands can be defined as those that have not been constructed by human activities.

47. A relatively new approach to wetland management is the construction of natural structures to treat all forms of water ___________________.

48. List the government and private agencies concerned with wetland management.

    a.

    b.

    c.

    d.

    e.

49. A new program allows destruction (grading, draining, filling, etc.) of one-half ___________________ of wetland without an individual permit.

50. Landowners are required to notify the U.S. Army Corps of Engineers of activities that impact any wetland area one- ___________________ of an acre or larger in size.

51. List the causes of wetland loss:

    a.

    b.

    c.

    d.

    e.

    f.

    g.

    h.

    i.

    j.

    k.

52. The best way to protect a wetland area is to provide a ___________________ zone around the wetland.

53. Detail the part restoration plays in the management of degraded wetlands, as mentioned in the chapter.

a.

b.

c.

54. List the government programs that support wetlands.

a.

b.

c.

d.

e.

f.

g.

h.

i.

j.

k.

l.

55. Most states direct ____________________ programs that help the public understand the value and function of the wetlands.

ACTIVITY 1

Purpose:

Evaluate a wetland habitat.

Research:

Review the information about wetlands detailed in the textbook.

Procedure:

Evaluate a local wetland found in your area.

Wetland Evaluation Data Form

Wetland name: Location:

Category	Condition	
	Good	**Poor**
Water depth		
Vegetation species		
Wildlife species		
Land use of area		
Human-made alteration to site		
Contamination of site		
Aesthetic value of site		
Water source of site		
Pollution controls at site		
Underground recharge capability		
Recreational opportunities		

Observations:

1. List the unique characteristics of the wetland.
2. What are the educational values of the wetland?

ACTIVITY 2

Purpose:

Describe the purpose of the Natural Resources Conservation Districts.

Research:

Locate the following information, provided by the instructor, about conservation districts: their origin, development, and functions.

Procedure:

Write a paragraph describing the role of the Natural Resources Conservation Districts.

Observations:

1. How might the Natural Resources Conservation District in your area assist you?
2. How might you become involved with your local Natural Resources Conservation District?

Land-Use Planning

TEST YOUR KNOWLEDGE

Complete the following:

1. __________ planning refers to the planning that is done for an individual farm or ranch.
2. The governing body in each locality establishes its set of __________ and regulations concerning how the land within its jurisdiction can be used.
3. The most common result of such a land use decision-making process is a set of "__________ regulations."
4. __________ regulations force like-activities to be grouped together.
5. What three things should land-use planning consider?

   a.

   b.

   c.
6. Increased rainfall __________ from parking lots can overtax storm drainage systems, and cause excessive erosion and sedimentation.
7. One of the most important farmer's responsibilities is __________ of the land.
8. The farmer is __________ of much of our nation's most important natural resources—our land with its thin covering of soil.
9. List four of the responsibilities a farmer has, as mentioned in the chapter.

   a.

   b.

   c.

   d.
10. The first requirement of any business is to make a __________.
11. The farmer must earn enough __________ from the farming enterprise to pay for the costs of operation and to meet the immediate needs of the farm family.

12. Business decisions cannot always be based on the __________ good of the farm.
13. Farming for __________, at times, takes preference.
14. List the reasons farming operations must produce income.

    a.

    b.

    c.

    d.

    e.

    f.

    g.

    h.

15. The individual farm's __________ program is the single most critical part of the soil management effort.
16. The first step in developing the farm __________ is to know the farm's physical features.
17. The soil __________ provided by the Natural Resources Conservation Service (NRCS) can be quite valuable.
18. With the help of the local soil conservationists, the cropping system can be __________ to the soil's capability to produce.
19. The basic principle is to match the crops to be grown to the soil's __________.
20. The better the land, the __________ row crops can be safely produced.
21. The United States has a total of __________ billion acres of land.
22. Cities, towns, and other housing areas take up __________ million acres in the United States.
23. Of our total of nearly 2.3 billion acres of land, only __________ billion is suitable for farming.
24. Considering only suitable lands, farmland is being used in the following manner:

    a.

    b.

    c.

    d.

25. Each year __________ million acres of prime farmland and 2 million acres of lesser-quality farmland are taken for nonfarm use.
26. Good __________ is the basis of our nation's strength and wealth.
27. Once farmland is covered with __________, it is lost to farming.
28. __________ or industrial land use is usually more profitable per acre than is farm use.
29. Land naturally tends to be put to its most __________ use.
30. As the amount of prime farmland decreases because of conversion, __________ productive land will be converted to farm use.
31. Once a field is __________ with asphalt or concrete, it is lost for food production.

ACTIVITY

Purpose:

Identify the symbols used on soil conservation maps.

Research:

1. Refer to the Figure below:

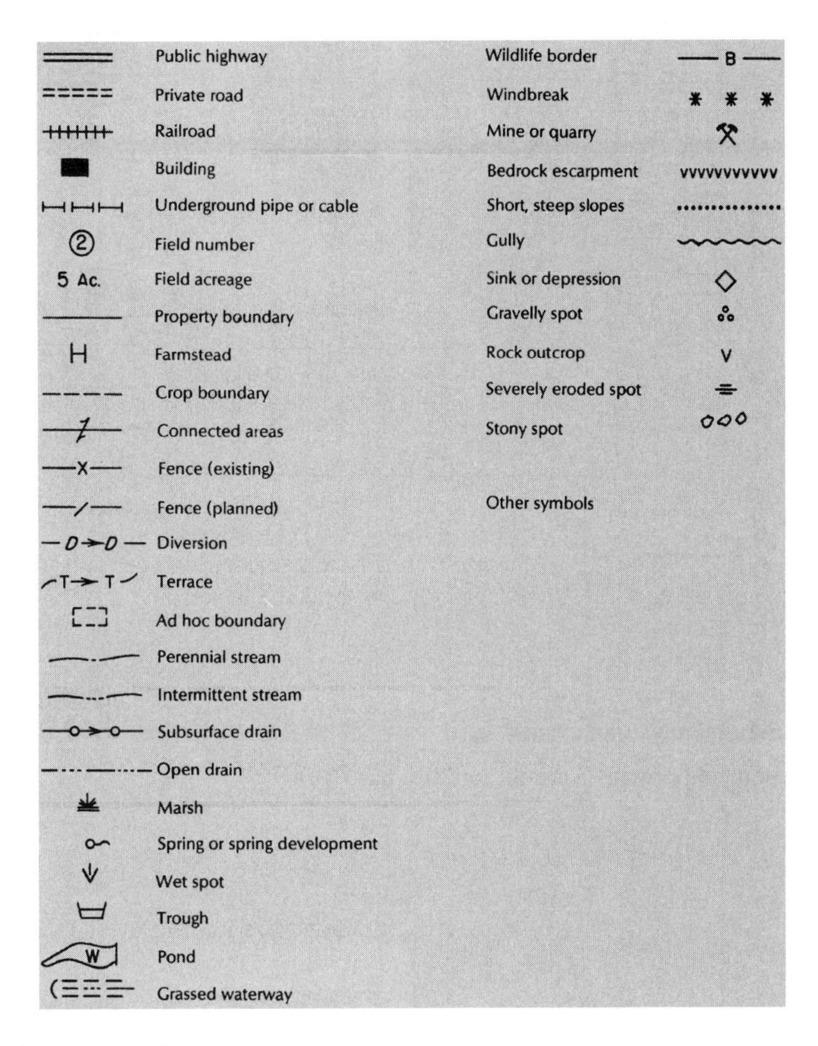

2. Review the given symbols.

Procedure:

Identify the following symbols:

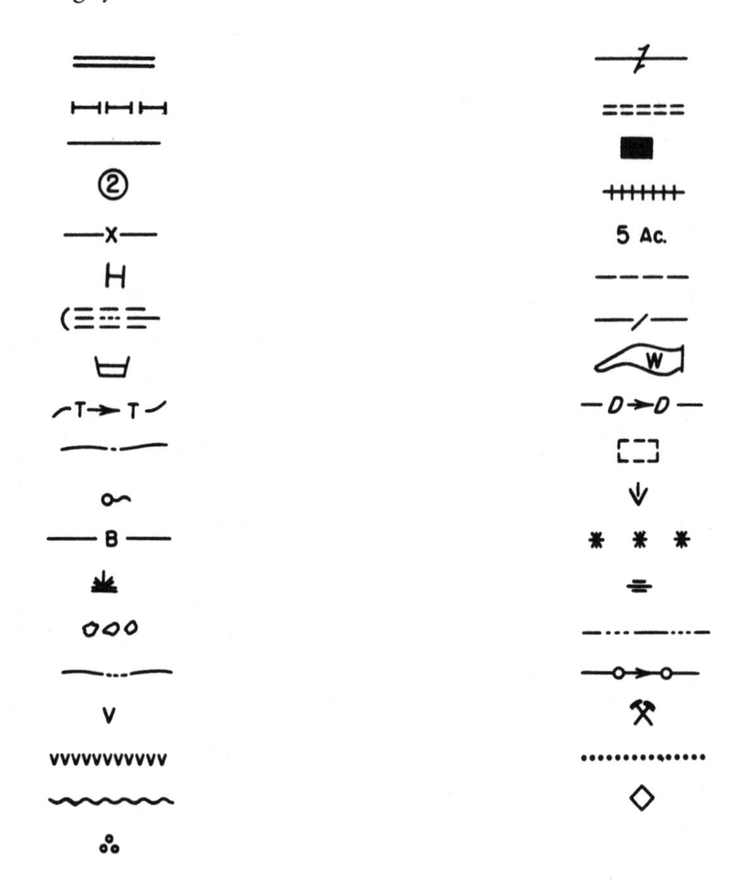

Observations:

1. Why are map symbols important?
2. What are some other map symbols that you have seen?

CHAPTER 12

Careers in Soil Management

TEST YOUR KNOWLEDGE

Complete the following:

1. Soil conservation workers are the primary group employed in occupations involving soil __________.
2. List the careers that make up the "other" groups involved in soil management.
   a.
   b.
   c.
   d.
3. Soil conservation workers are employed by:
   a.
   b.
   c.
   d.
4. Soil __________ provide advice and assistance to farmers, ranchers, and others.
5. Soil conservationists help to develop farm __________ that take the greatest advantage of the ranch or farmland's capability.
6. __________ managers specialize in conservation in the grasslands.
7. Range managers or range __________ help ranchers, farmers, other landowners, and public officials in the range areas plan for the wise use of soil and water resources.
8. Careful range __________ planning is important.
9. If erosion is a problem, the __________ is found and a plan to correct it is developed.

10. List a soil conservationist's responsibilities.

    a.

    b.

    c.

    d.

    e.

    f.

    g.

    h.

    i.

    j.

11. __________ is the ability of the soil to allow water to move through it.
12. A __________ degree is a minimum for entry into soil conservation work.
13. The range manager or soil conservationist must be able to __________ well with the public.
14. Public __________ is convincing the people in the area that soil and water conservation and range management are important.
15. __________ interviews and newspaper interviews are common requests of a range manager or soil conservationist.
16. The soil conservation __________ provides technical assistance on the soil or other related matters to the soil conservationist.
17. A soil conservation technician must be able to operate surveying equipment, __________ levels, and other measuring devices.
18. The soil conservation technician helps in the supervision of soil conservation __________.
19. High school graduation is required for entry into the career of soil conservation __________.
20. The soil __________ studies the physical, chemical, biological, and mineralogical composition of the soil.
21. Soil scientists use aerial __________, topographic maps, and the information derived to develop soil maps.
22. A soil __________ is a collection of soil type maps and other related information.
23. The soil scientist develops the technical and scientific __________ that makes possible much of the fieldwork of soil conservation and land-use planning.
24. Research soil scientists conduct __________ on fertilizers, soil chemistry, physics, and biology.
25. Work as a research soil scientist will require both masters and __________ degrees.
26. The soil scientist must be able to examine a problem, apply theoretical reasoning and experimentation, and arrive at a logical __________.
27. A soils __________ analyzes and evaluates soils for construction sites, ponds and lake sites, erosion-control construction, and other conservation activities.
28. A soil engineer must evaluate:

    a.

    b.

    c.

    d.

    e.

29. Percolation rate refers to the __________ at which water soaks into the ground.
30. To become a soils engineer, a degree in mechanical or civil engineering or in __________ engineering is required.
31. The number of soils engineers to be employed is expected to __________ by 22 percent through 2020.

ACTIVITY

Purpose:

Determine the length of pace.

Materials Needed:

100-foot tape

Research:

Pacing is one method of measuring distance. It consists of determining the average length of normal steps or paces. The length of a pace varies with the individual, the rate of speed, and the terrain.

Procedure:

1. Measure a distance of 100 feet.
2. Pace its length at least four times.
3. Divide the number of paces by four to determine the average number of paces in 100 feet.
4. Divide 100 feet by the "average number of paces."

Practice pace	Number of paces
1	
2	
3	
4	
Total	

Observations:

How many feet is your pace?

$$\frac{\text{Total number of paces}}{4} = \text{Average number of paces}$$

$$\frac{\text{100 feet}}{\text{(average number of paces)}} = \frac{\text{feet}}{\text{pace}}$$

JOB EXERCISE FOR CHAPTER 4–12

Purpose:

Classify groups of rocks.

Materials Needed:

Rock and Mineral Kit
Rock and Mineral Study Guide

Research:

By using the materials, learn to identify the rocks listed here:

Conglomerate
Serpentinite
Marble
Limestone
Pumice
Obsidian
Slate
Anthracite Coal
Calcareous Tufa
Shale
Basalt
Talc Schist
Scoria
Sandstone
Granite

Procedure:

1. Define igneous rock.
2. Define sedimentary rock.
3. Define metamorphic rock.

Observations:

1. Determine what group each rock belongs to.
2. Complete the table by placing a check mark in the appropriate box.

Rock	Igneous	Sedimentary	Metamorphic
Conglomerate			
Serpentinite			
Marble			
Limestone			
Pumice			
Obsidian			
Slate			
Anthracite Coal			
Calcareous Tufa			
Shale			
Basalt			
Talc Schist			
Scoria			
Sandstone			
Granite			

UNIT

III

Water and Air Resources

CHAPTER

13

Water Supply and Water Users

TEST YOUR KNOWLEDGE

Complete the following:

1. Water is a necessary ingredient for all __________ organisms.
2. Water covers about __________ % of the earth's surface.
3. __________ % of the earth's water is located in the oceans and seas, 2% is freshwater, and 1% is frozen in glaciers.
4. The amount of water on this planet is fairly __________.
5. The __________ of water is not constant.
6. Water is continually moving from place to place by means of the water cycle, or__________ cycle.
7. The movement of water into the atmosphere is called __________.
8. Water will eventually return to the ocean in the form of __________, sleet, or snow.
9. Returned water to the ocean is called __________ water.
10. Plants absorb water through their roots and release it through tiny openings in their leaves called __________.
11. The evaporation of water from plants is called __________.
12. List the functions of water in animals.

    a.

    b.

    c.
13. __________ water is used by people over and over as it makes its way toward the ocean.
14. If the surface __________ water is too great, it can be harmful.
15. Water that soaks into the soil adds to the __________ supply.
16. The zone of __________ is made up of empty spaces, filled with air, between soil mineral particles.
17. Below the zone of aeration is the groundwater zone where water is trapped in water-saturated rocks called __________.

18. The main components of the hydrologic cycle include:

    a.

    b.

    c.

    d.

    e.

19. __________ gallons of water equals 1 cubic foot.
20. About __________ cubic miles of meteoric water falls in the United States each year.
21. The United States withdraws from the water cycle about __________ cubic miles of water.
22. Domestic uses account for about __________% of U.S. water use.
23. Industrial uses account for __________ % of U.S. water use.
24. __________ % of U.S. water use is for agricultural use.
25. Almost all water withdrawn from the hydrologic cycle eventually __________ to the cycle.
26. The U.S. population drinks about __________ million gallons of water per day.
27. The main users of water include:

    a.

    b.

    c.

    d.

    e.

    f.

28. The main agricultural use of water is __________.
29. The most common methods of irrigation include surface and __________ techniques.
30. __________ irrigation involves building a series of large and small ditches to transport water.
31. The surface irrigation technique can use either a flood or a __________ method.
32. The __________ irrigation method involves flooding with a continuous sheet of water on a strip down the field.
33. The __________ irrigation method uses land furrows to move the water to the field.
34. The __________ irrigation system involves applying water over the top of crops.
35. The most common method of __________ irrigation sprinkler system is to supply water to a lateral that pivots around the supply.
36. The __________ irrigation sprinkler system involves burying underground waterlines in the area to be irrigated.
37. __________ irrigation consists of water supply pipes with lateral tubes going to individual plants.
38. A __________ is attached to the end of the drip irrigation tube.
39. __________ are rated by the amount of water (gallons per hour) they supply to the plants.
40. The continual sinking of vast __________ to tap the water supply is creating a drain on that supply.
41. Irrigation water not drawn from groundwater it must be drawn from __________ water.
42. When irrigation water is lost to evaporation, mineral and __________ content intensifies.
43. The high concentration of irrigation salts in the water is termed __________.
44. Industry uses more __________ than any other raw material.
45. Most industry-used water is either __________ for another use or returned to the natural water cycle.

46. Industry draws about __________ billion gallons of water per day.
47. A major use of water is in the production of __________ from water-driven turbine generators.
48. The United States leads the world in __________ power.
49. The principle behind the production of electricity from waterpower involves tapping the __________ from water moving from a high place.
50. The way that water is __________ affects not only the fish in the water but also all the wildlife around the water.
51. List the recreational activities that water provides, as mentioned in the chapter.

    a.

    b.

    c.

    d.

    e.

52. Each person in the United States uses about __________ gallons of water per day.

ACTIVITY

Purpose:

Identify surface water supplies.

Research:

1. Locate a state map.
2. Locate the following surface water areas in your area detailed on the map:
   a. Lakes
   b. Ponds
   c. Rivers
   d. Reservoirs
   e. Streams or creeks

Procedure:

Complete the following table by listing the names of the surface water in your area.

Lakes	Rivers	Streams	Ponds	Reservoirs

Observations:

1. What is the main source of surface water supplies in your area?
2. Considering the main source of surface water, what is the name of the major supplier of this surface water?

CHAPTER

14

Water Pollution

TEST YOUR KNOWLEDGE

Complete the following:

1. For thousands of years, we regarded water as a means of disposing of __________.
2. Three basic sources of water pollution include:

    a.

    b.

    c.
3. __________ source pollution results from the direct introduction of contaminants into the water supply at an identifiable location or multiple locations.
4. When the source of contamination can be precisely located, we refer to it as __________ source pollution.
5. __________ source pollution is referred to as nonpoint source pollution.
6. Diffuse pollution results from the introduction of contaminants across a __________ area.
7. Diffuse pollution can be somewhat localized but cannot be tied to a __________ input location.
8. __________ pollution results from ongoing, natural processes.
9. Water is one of nature's most universal __________.
10. As water __________, it generates forces that pick up particles and carry them along.
11. Many minerals and rocks are __________.
12. Ongoing contamination of the water supply from natural sources is termed __________ pollution.
13. __________were common in the United States until the middle of the 1900s.
14. As populations grew, cities had to install sewage __________ systems to take care of the large amounts of wastes.
15. __________ pollution can be anything from detergents used to clean the roads to salts used to melt the snow and ice in the winter.
16. Pollution from cities affects not only the surface water runoff but also the __________ supply.
17. Municipal dumps and __________ are potential sources of groundwater pollution.

18. Today, pollution of groundwater supplies from landfills is being monitored by __________ wells close to the landfill sites.
19. Because industry has more to lose from water pollution than any other segment of our economy, industry has also done the __________ to control water pollution.
20. List the four major categories that industrial pollution of water can be divided into.

    a.

    b.

    c.

    d.
21. __________ pollution involves returning heated water to a stream or river.
22. The Federal Water Pollution Control Administration has attributed the largest industrial thermal pollution to:

    a.

    b.

    c.
23. Whenever energy is changed from one form to another, __________ is lost.
24. The dumping of heated water into a stream can create problems for both the stream or lake and the __________ in and around the water.
25. Some fish use water temperature to trigger __________ and migration.
26. Warmed water will contain less __________, thus limiting the fish population the water can support.
27. If the heated water is discharged for a long period of time, increased growth of __________ will start to choke the stream.
28. If algae growth becomes substantial, the algae will emit __________ that will eventually kill the fish population in the stream.
29. Techniques discussed in the chapter, used to control thermal pollution are:

    a.

    b.
30. In cooling __________, the water passes through a series of coils around which air is forced by a fan, and the moving air cools the water before it is returned to the stream.
31. A cooling __________ is a pond beside the plant into which heated water flows.
32. __________ material elements emit radiation as a result of the disintegration of their atomic nuclei.
33. Fish and wildlife accumulate radioactive materials in their __________.
34. The only way radioactive materials are destroyed is by the passage of __________.
35. If waters are polluted with organic wastes, bacteria will grow rapidly, placing great demands on the __________ in the water.
36. If oxygen is not available to fish because of high bacteria content, the fish will actually __________ and die.
37. Scientists have developed the biological oxygen demand test, referred to as the __________ test to determine the oxygen demand on water that directly indicates the amount of wastes in the water.
38. __________ industrial wastes are common in the manufacture of soaps and detergents, drugs and pharmaceuticals, paints, and fertilizers.
39. The need to produce more and more food on less and less land has led to the development of __________ farming practices.

40. The most common agricultural pollutants include:

    a.

    b.

    c.

    d.

41. Intense livestock areas create problems in the __________ of the waste materials.
42. If the livestock wastes are not properly __________ of, both surface and groundwater will be affected.
43. If agricultural chemicals and fertilizers are correctly used, they normally __________ to the soil particles and stay there.
44. When rain comes, extra __________ chemicals may either run off into a stream or leach into the groundwater supply.
45. When great amounts of fertilizer, especially nitrates and phosphates, reach the water in ponds or streams, the water becomes __________ rich.
46. Excessive enrichment of water is called __________.
47. Erosion causes topsoil to move into nearby lakes and __________.
48. Once the water slows down the soil will __________ to the bottom of the lake or stream.
49. The principal problem of water pollution control is to determine how much pollution is __________.
50. One method of detecting water pollution involves examining streams for populations of __________.
51. Invertebrates are animals having no __________ columns.
52. The invertebrates used in detecting water pollution are __________, which are large enough to have complex bodies and are visible to the naked eye.
53. Examples of pollution-intolerant macroinvertebrates include:

    a.

    b.

    c.

    d.

    e.

    f.

54. Examples of organisms that are somewhat pollution-tolerant are:

    a.

    b.

    c.

    d.

    e.

    f.

    g.

    h.

    i.

    j.

    k.

    l.

55. Examples of organisms that are pollution-tolerant are:

    a.

    b.

    c.

    d.

    e.

    f.

56. By identifying the macroinvertebrates present in a stream or lake, you can roughly determine the water __________ in the stream.

57. If only pollution-tolerant organisms are present, the water quality is __________.

58. If a larger number of pollution-intolerant organisms are present, the water quality is probably __________.

ACTIVITY

Purpose:

Determine pollution sources.

Research:

Review the three major water pollution groups. Water pollution can be separated by its source into two categories—point-source and nonpoint-source pollution. Point-source pollution includes materials that are discharged directly from a specific source. Nonpoint-source includes runoff from sources that are widespread and harder to identify.

Procedure:

Complete the following table using the following information:

Source:
PS — Point source
NP — Nonpoint source (Diffuse)

Pollution Group:
U — Urban
I — Industrial
A — Agricultural

Pollution	Source	Group
Detergents used to clean roads		
Sediments from fields		
Radioactive waste from a medical lab		
Salts to melt snow on roads		
Heated water from a power plant		
Feedlot waste		
Fertilizer applied to fields		
Pesticide residue from fields		
Radioactive waste from a nuclear power plant		
Landfill		
Waste from production of detergent		
Municipal dump		
Waste from production of fertilizer		
Municipal sewage treatment discharge		
Failing septic tanks		
Abandoned mine drainage		
Ocean dumping of city wastes		

Observations:

1. Of the types of pollution listed, which source occurred most often?
2. Of the types of pollution listed, which group occurred most often?

CHAPTER

15

Water Purification and Wastewater Treatment

TEST YOUR KNOWLEDGE

Complete the following:

1. __________ is one of the most basic natural resources.
2. To thrive as an organized society, humans must remove significant amounts of water, use it in many different ways, and return almost all of it to the __________ cycle.
3. In the entire world, there are about __________ cubic miles of water.
4. Humans use about __________ cubic miles of water annually.
5. Use means that we remove water from the natural water cycle, use it, and then __________ almost all of it to the natural water cycle.
6. Almost all human activity relies on a reliable, abundant supply of __________ water.
7. Human waste products can come in the form of:

   a.

   b.

   c.
8. Water from clean __________ needs little pretreatment before it is suitable for use.
9. Water from municipal reservoirs, upland streams, and lakes with limited inflow may need only __________ treatment.
10. Water from rivers into which industrial, agricultural, and municipal wastes have been introduced upstream may require __________ treatment.
11. Once the water has been used, it must be __________ before it can be safely returned to the hydrologic cycle.
12. List the two basic wastewater treatment systems.

    a.

    b.

13. Individual household wastewater treatment systems consist almost exclusively of ____________ treatment systems.
14. In nature water is never ____________ and clean.
15. ____________ is one of the most effective of all solvents.
16. ____________ are substances that dissolve other substances.
17. When one substance is dissolved in another, it is said to be a __________.
18. In a solution, the __________ of one substance are dissipated among the molecules of another substance.
19. Water is never pure in nature because it holds many other substances in __________.
20. Large particles may be held in suspension only while the water is moving very __________.
21. Very __________ particles may remain in suspension for extended periods of time.
22. Before we use water for human consumption, __________ must be removed and many naturally occurring conditions must be changed.
23. List the three broad categories of impurities removed during the treatment process.
    a.
    b.
    c.
24. We think of "pure" water as water that has no color, no taste, and no __________.
25. Anything that causes a strange taste, color, or odor in water is __________ undesirable.
26. Chemical impurities can result in these three conditions, as mentioned in the chapter.
    a.
    b.
    c.
27. To be usable for human consumption, water should be neither very acid nor very __________.
28. The pH of water should be near __________ neutral.
29. Water with a pH much lower than 7.0 is __________ and can produce corrosive effects on metal equipment and pipes.
30. Water with a pH higher than 7.0 is __________.
31. Very alkaline water is not only __________ but also tends to produce scale.
32. __________ water comes as close to being pure water as most people will ever encounter.
33. Low levels of dissolved __________ and metals in drinking water are usually beneficial for a person's health.
34. Water from ground supplies often contains dissolved __________ or manganese.
35. When water with iron and manganese is exposed to air, the metals __________, causing red or brown stains.
36. If the concentration of metals in water is very high, the water will have a __________ color and even smell like rust.
37. __________ water is a result of excess calcium or magnesium.
38. Calcium carbonate and magnesium hydroxide both precipitate out to form scale deposits that __________ water pipes and plumbing fixtures.
39. When sulfur is exposed to oxygen in water, it forms __________ dioxide gas.
40. Sulfur dioxide creates the odor that results when eggs begin to __________.
41. Excessive nitrogen and phosphorus promote the growth of __________ in water.

42. List the dissolved hydrocarbons that cause problems in water:

    a.

    b.

    c.

    d.

    e.

    f.

43. __________ impurities in freshwater that will be used for human consumption must be removed or neutralized.

44. In water treatment, fish, crustaceans, worms, and macroinvertebrates are removed at the outset by __________.

45. __________ are tiny green plants that grow in sunlight and in the presence of air.

46. Algae in excessive concentrations can extract almost all __________ from water so that other organisms such as fish are killed.

47. Bacteria and fungi can serve as decomposers and break down __________ matter and chemicals into harmless compounds.

48. __________ are disease-causing organisms.

49. __________ may become pathogens when humans or animals drink the water, or even when they are exposed to water.

50. Common bacterial diseases carried in water are __________ fever and cholera.

51. __________ are single-celled animals that occur naturally in water.

52. Common human diseases caused by protozoa are __________ and dysentery.

53. __________ refers to solid matter suspended in a liquid.

54. All naturally occurring water has at least some degree of __________.

55. When the concentration of solids in suspension becomes great enough to be visible, the water is __________ unpleasing.

56. The objective of water treatment is to produce a __________ water supply.

57. __________ water is water that is chemically and microbiologically safe, and that is otherwise suitable for human consumption.

58. Generally, __________ water is safe to drink, as it comes from the ground.

59. Well water should be __________ periodically by local authorities for biological or chemical contaminants to ensure that it is safe to drink.

60. Well water may contain undesirable levels of chemicals such as:

    a.

    b.

    c.

    d.

    e.

61. List household treatment systems, discussed in the chapter.

    a.

    b.

    c.

62. Water that is moving into a treatment system is referred to as __________.
63. Water that comes out the other end of the treatment system and is ready for use is called __________.
64. Influent is first __________ to remove the largest particles of solid matter and large organisms.
65. A __________ is a chemical that promotes the coagulation of solid materials in suspension.
66. ____________ is the physical process of smaller particles clumping together to form larger particles that will later be allowed to settle out of the water.
67. The water/coagulator mix passes into a __________.
68. The __________ is a large tank in which paddles gently stir the water to encourage the clumping together of large particles of solid matter that will settle out later.
69. The coagulated, chemically treated water flows out of the flocculator into the bottom of a vertical __________ tank.
70. A very gentle __________ system mechanically continually removes the accumulated settlings from the bottom of the tank.
71. The water is removed from the __________ of the settling tank and passed through a final filter made up of graduated layers of sand and gravel, and still more activated carbon.
72. From the final filter water will be treated with __________ and fluorine and moved into the distribution system for disposal or use.
73. Once the water leaves the settling tank, it is passed through an __________ carbon filter that starts to neutralize odors and tastes.
74. The basic form of human sewage disposal is through the use of a __________ system.
75. List the four parts a typical septic system consists of.

    a.

    b.

    c.

    d.

76. Wastewater treatment is divided into three phases:

    a.

    b.

    c.

77. The primary waste management system removes about __________-thirds of the wastes from water.
78. The secondary waste treatment system involves __________ processing of sewage.
79. The tertiary waste treatment system is the __________ processing of sewage wastewater.
80. The sewage treatment __________ can be beneficial as a soil builder and fertilizer.

ACTIVITY

Purpose:

Interpret waste analysis results.

Research:

Review the standards detailed here:

Standards	+	+
Total Suspended Solids		less than 45 mg/l
Biological Oxygen Demand		less than 45 mg/l
Ammonia		less than 12.75 mg/l
pH		6.5 to 8.3
Dissolved Oxygen		more than 6.0
Total Residual Chlorine		less than 0.13 mg/l

Water released from the wastewater treatment plant must not exceed these standards.

Procedure:

Complete the table by indicating which of the following samples are not safe for release into the stream.

Standards

Information	Sample #1	Sample #2	Sample #3
Total Suspended Solids	50 mg/l	16 mg/l	25 mg/l
Biological Oxygen Demand	50 mg/l	13 mg/l	17 mg/l
Ammonia	2.67 mg/l	1.57 mg/l	15 mg/l
pH	7.3	7.34	8.1
Dissolved Oxygen	6.0 mg/l	7.38 mg/l	6.0 mg/l
Total Residual Chlorine	0.07 mg/l	0.02 mg/l	0.07 mg/l

Observations:

List the sample number that does not meet the standards and indicate which standards are not met.

Sample	Standards	

CHAPTER 16

Water-Use Planning

TEST YOUR KNOWLEDGE

Complete the following:

1. The principal water management techniques discussed in this chapter include:
   a.
   b.
   c.
   d.
   e.
   f.
2. Dams, reservoirs, and ponds have what basic uses?
   a.
   b.
   c.
   d.
   e.
   f.
   g.
3. It is usually the goal of the water manager to ____________ flood control and fish populations.
4. The life of the reservoirs or dams will be greatly reduced if ____________ resulting from soil erosion is allowed to pour into the basin.
5. Another water management use is the use of water to transport ____________.
6. The power to move goods over water is ____________ than that required to move them over land.
7. The most common water-transported items are ____________ and coal products.
8. The largest bodies of water are the ____________.

9. Owing to its high __________ content, ocean water is not useful in its present state.
10. The process of salt extraction is called __________.
11. More than 97% of the world's water has salinity that __________ safe drinking standards.
12. The World Health Organization specifies that total dissolved solids (TDS) not exceed __________ mg/l.
13. Seawater averages __________ mg/l in TDS.
14. The process of __________ has become critical in supplying the needs of many parts of the world for fresh water.
15. The main goals of the Office of Saline Water (OSW) are:

    a.

    b.

    c.

16. Methods of salt removal are:

    a.

    b.

    c.

    d.

    e.

17. The __________ process of desalination is used mainly on inland brackish water where the salt content is lower.
18. The membranes are __________ that allow water to pass through but keep the salts out.
19. The main methods of membrane desalination are:

    a.

    b.

20. __________ osmosis uses the natural principle of liquids flowing through a semipermeable membrane.
21. Reverse osmosis applies __________ to the saltwater to make it flow through a membrane.
22. When subjected to an electric current, sodium and chlorine separate and pass through different filters in the __________ process.
23. The __________ process works when water is heated, then evaporates, and leaves behind any solid impurities.
24. The three main distillation processes developed are:

    a.

    b.

    c.

25. In __________ distillation, saltwater is sent through tubes where it is heated by steam.
26. The __________ compression distillation process involves placing the water under pressure when turned into steam.
27. __________ distillation uses the fact that water boils at a lower temperature when it is at a lower pressure.
28. In a typical home, an average person uses __________ –75 gallons of water daily.

29. List the uses of water in households:

    a.

    b.

    c.

    d.

    e.

30. The most common and oldest method of making rain is to seed the clouds with silver ___________ crystals.
31. The problem with water is that in so many cases, the water is not where the ___________ are.
32. A 1-inch rain dumps ___________ gallons of water on each acre of land.
33. The main problem facing experts in recycling is the removal of ___________, tastes, and salts from the water.

ACTIVITY

Purpose:

Calculate residential water rates.

Research:

Residential and commercial water rates differ because of the amount of water used, as shown:

$$0 - 300 \text{ cubic feet} = \frac{\$1.00}{100 \text{ cubic feet}}$$

$$300 + \text{ cubic feet} = \frac{\$0.75}{100 \text{ cubic feet}}$$

Water rates decline as the total amount of water increases. Residential water bills may include garbage and sewer costs.

Review this example:

Consider a present reading of 15,000 and a previous reading of 10,000:

Present reading	~150 (100 cubic feet)
Previous reading	~100 (100 cubic feet)
	~50 (100 cubic feet)

$$3 \times 1.00 = \$3.00$$

$$50 - 3 = 47 \times 0.75 = \$35.25$$

$$= \$38.25$$

Procedure:

Complete the following billing statements using the prices detailed in the Research section. Calculate: 100 cubic feet used, amount billed, total due.

① **MUNICIPAL UTILITY BILL**

CITY FINANCE OFFICE — UTILITY BILL DEPT.
222 MAIN ST. — SOMEWHERE, U.S.A. 12345
321-1234

SERVICE AT	714 WRIGHT CRT			
METER READING PRESENT	METER READING PREVIOUS	100 CU. FT. USED	CODE	AMOUNT BILLED
149	117		WC	
METER READING DATES			GC	4.00
073100	070100		SC	7.70

BILLS PAI AFTER DUE DATE SHOWN BELOW ARE SUBJECT TO A LATE PAYMENT CHARGE

DESCRIPTION OF CODES ON REVERSE SIDE

BALANCE FORWARD	AMOUNT BILLED	TOTAL DUE
ACCOUNT NUMBER	BILLING DATE	DUE DATE
033199006	08/28/00	09/22/00
LAST PMT. DATE	TOTAL PMTS. REC'D	LAST ADJ.
08/17/00	35.90	

KEEP THIS STUB FOR YOUR RECORDS

② **MUNICIPAL UTILITY BILL**

CITY FINANCE OFFICE — UTILITY BILL DEPT.
222 MAIN ST. — SOMEWHERE, U.S.A. 12345
321-1234

SERVICE AT	714 WRIGHT CRT			
METER READING PRESENT	METER READING PREVIOUS	100 CU. FT. USED	CODE	AMOUNT BILLED
149	100		WC	
METER READING DATES			GC	4.00
073100	070100		SC	7.70

BILLS PAID AFTER DUE DATE SHOWN BELOW ARE SUBJECT TO A LATE PAYMENT CHARGE

DESCRIPTION OF CODES ON REVERSE SIDE

BALANCE FORWARD	AMOUNT BILLED	TOTAL DUE
ACCOUNT NUMBER	BILLING DATE	DUE DATE
033199006	08/28/00	09/22/00
LAST PMT. DATE	TOTAL PMTS. REC'D	LAST ADJ.
08/17/00	35.90	

KEEP THIS STUB FOR YOUR RECORDS

③ **MUNICIPAL UTILITY BILL**

CITY FINANCE OFFICE — UTILITY BILL DEPT.
222 MAIN ST. — SOMEWHERE, U.S.A. 12345
321-1234

SERVICE AT	714 WRIGHT CRT			
METER READING PRESENT	METER READING PREVIOUS	100 CU. FT. USED	CODE	AMOUNT BILLED
345	75		WC	
METER READING DATES			GC	4.00
073100	070100		SC	7.70

BILLS PAID AFTER DUE DATE SHOWN BELOW ARE SUBJECT TO A LATE PAYMENT CHARGE

DESCRIPTION OF CODES ON REVERSE SIDE

BALANCE FORWARD	AMOUNT BILLED	TOTAL DUE
ACCOUNT NUMBER	BILLING DATE	DUE DATE
033199006	08/28/00	09/22/00
LAST PMT. DATE	TOTAL PMTS. REC'D	LAST ADJ.
08/17/00	35.90	

KEEP THIS STUB FOR YOUR RECORDS

④ **MUNICIPAL UTILITY BILL**

CITY FINANCE OFFICE — UTILITY BILL DEPT.
222 MAIN ST. — SOMEWHERE, U.S.A. 12345
321-1234

SERVICE AT	714 WRIGHT CRT			
METER READING PRESENT	METER READING PREVIOUS	100 CU. FT. USED	CODE	AMOUNT BILLED
555	65		WC	
METER READING DATES			GC	4.00
073100	070100		SC	7.70

BILLS PAID AFTER DUE DATE SHOWN BELOW ARE SUBJECT TO A LATE PAYMENT CHARGE

DESCRIPTION OF CODES ON REVERSE SIDE

BALANCE FORWARD	AMOUNT BILLED	TOTAL DUE
ACCOUNT NUMBER	BILLING DATE	DUE DATE
033199006	08/28/00	09/22/00
LAST PMT. DATE	TOTAL PMTS. REC'D	LAST ADJ.
08/17/00	35.90	

KEEP THIS STUB FOR YOUR RECORDS

CHAPTER

17

Air and Air Quality

TEST YOUR KNOWLEDGE

Complete the following:

1. The blanket of gasses that surround the earth is generally referred to as the __________.
2. The atmosphere reaches from the earth's surface to a height of about __________ miles above sea level.
3. __________ level is defined as the mean height of the ocean between high tide and low tide.
4. Sea level is defined as having an elevation of __________ and all other elevations are measured in terms of height "above sea level."
5. Scientists divide the atmosphere into what zones based on height above sea level?

   a.

   b.

   c.

   d.

   e.
6. The __________ starts at the surface and extends upward to an average of about 4 miles over the poles and ranges up to 17 miles above sea level near the equator.
7. The troposphere is the region of the atmosphere in which temperatures __________ as the distance from the surface increases.
8. In the __________, as the elevation increases, the temperature also increases because the atmospheric gasses absorb so much of the sun's ultraviolet light energy as heat.
9. The majority of atmospheric gas is __________ (N_2), accounting for about 78% of the atmosphere.
10. __________ gas (O_2) makes up 21% of the atmosphere.
11. The atmosphere holds a tremendous amount of __________ in the form of water vapor, water droplets, and ice crystals.
12. __________ matter are pieces of solid material or liquid suspended in the atmosphere.
13. __________ is used to describe things that result from human activity.

14. Gasses and particulates that result from human activity are known as ___________ emissions.
15. The earth's atmosphere is about ___________% nitrogen and oxygen.
16. List the two reasons to be concerned about nitrogen and oxygen.

    a.

    b.

17. Scientists believe that the increase in greenhouse gasses is contributing to ___________ warming.
18. Carbon ___________ (CO_2) is a colorless, odorless, tasteless, nontoxic, noncombustible, naturally occurring gas that makes up about 4/100s of 1% of the atmosphere.
19. Carbon ___________ is a by-product of burning carbon-based materials like wood and oil.
20. As a liquid, CO_2 sold in steel drums and tankers is most commonly used as an ingredient in soft drinks, producing their "___________."
21. In the atmosphere, the concentration of CO_2 gas is measured in parts per ___________ (ppm).
22. Carbon ___________ (CO) is a colorless, odorless, and tasteless gas that will burn (producing CO_2).
23. In the atmosphere, the naturally occurring CO level is about ___________ ppm, which is not harmful to humans.
24. In higher concentrations, CO is highly ___________ to animals, including humans.
25. Volcanic eruptions and the incomplete natural ___________ of forests, bushes, and grasses produce CO gas.
26. Carbon monoxide levels in the atmosphere have been ___________ steadily for the past 30 years.
27. To avoid CO injury, it is essential that ___________ systems in homes and businesses be properly maintained.
28. Homes and businesses should be protected by carbon monoxide ___________.
29. ___________ oxides occur in the atmosphere naturally but anthropogenic generation takes place through burning fossil fuels, land conversion from forests to agriculture use, burning biomass, and the addition of nitrogen fertilizer to the soil.
30. List the combinations of nitrogen and oxygen.

    a.

    b.

    c.

    d.

    e.

    f.

31. ___________ (CFCs) are strictly an anthropogenic problem.
32. CFCs were developed in the 1930s for use in industry as ___________ and as propellants and cleaning agents.
33. Scientists discovered CFCs in the upper atmosphere and connected the compounds to the ___________ of ozone (O_3) gas.
34. CFCs are known to act as ___________ to break down ozone molecules into the more stable O_2 gas.
35. ___________ (O_3) is a form of oxygen gas containing three atoms of oxygen instead of two.
36. In the upper atmosphere, ozone is important in minimizing the amount of short-wave ___________ (UV) light that reaches the surface of the earth.
37. The so-called ozone ___________ is that part of the stratosphere with the highest ozone concentration.
38. UV light in large doses is ___________ to most living beings, so ozone is important because it filters out UV light.
39. The process of anthropogenic ozone loss in the stratosphere has come to be known as ozone ___________.

40. The seasonal decline in ozone concentrations in the stratosphere above the poles is often referred to as the ___________ hole.
41. ___________ (NH_4) is colorless, odorless, and flammable.
42. Methane occurs in natural gas and is formed when organic matter ___________.
43. Large quantities of methane are produced in the ___________ systems of animals, particularly ruminants such as cattle.
44. ___________ was a component in gasoline and used in large quantities to increase the efficiency of motor vehicle engines.
45. Lead was found to be a significant health hazard and banned from ___________ and paint in most parts of the world in the 1970s.
46. ___________ is very common in nature and introduced into the atmosphere most notably from natural fires and volcanic eruptions.
47. When sulfur enters the atmosphere, it combines with oxygen to produce sulfur ___________ (SO_2).
48. SO_2 is a major culprit in the generation of ___________ rain.
49. The ___________ effect is the process by which infrared radiation produces an increase in the observable temperature in a closed system.
50. Short-wave radiation passes through glass but when it strikes the surface inside the car, it is absorbed and produces long-wave energy, which we sense as ___________.
51. The greenhouse gasses in the atmosphere capture most of the ___________ energy.
52. As greenhouse gas concentrations have increased over the past two centuries, the greenhouse effect has gradually captured ___________ amounts of solar energy in the form of heat.
53. Scientists believe that part of the increase in the average global temperature is a result of ___________ emissions.
54. Carbon ___________ is the long-term storage of carbon in the biosphere, underground, and ocean.

ACTIVITY

Purpose:

Understand Air and Air Quality terms.

Procedure:

Write the definition of the following terms:

1. Troposphere
2. Stratosphere
3. Mesosphere
4. Thermosphere
5. Exosphere
6. Particulate matter
7. Anthropogenic
8. Anthropogenic emissions
9. Carbon dioxide
10. Carbon monoxide
11. Ozone depletion
12. Ozone holes
13. Methane
14. Greenhouse effect
15. Carbon sequestration

Observations:

Which of the above terms you have used in a recent conversation with your friends?

CHAPTER

18

Careers in Water and Air Management

TEST YOUR KNOWLEDGE

Complete the following:

1. ______________ in water management range from predicting when and where water will fall, to careers in finding out where it went.
2. As the requirement for ______________ decreases, so do the salaries received.
3. The four main environmental science occupations listed in this chapter are:
   a.
   b.
   c.
   d.
4. The U.S. Department of Labor groups geologists, geophysicists, and oceanographers together under ______________.
5. ______________ study the structure, composition, and history of the earth's surface.
6. Geologists use tools such as:
   a.
   b.
   c.
   d.
   e.
   f.
7. Geologists ______________ construction companies and governmental agencies on the suitability of constructing dams, buildings, and highways; manage research projects; or teach in universities or colleges.

8. A groundwater geologist studies the ____________ and reserves of underground water resources.
9. Many times the geologist works____________.
10. A bachelor's degree in ____________ or a related field such as earth science is a minimum for a few entry-level positions.
11. Geologists usually begin their career in field exploration or as research assistants in ____________.
12. According to the U.S. Department of Labor, geoscientists who can speak a foreign ____________ and are willing to travel abroad will have some of the best opportunities.
13. ____________ is a catchall term of the scientific study of the composition and physical aspect of the earth.
14. Geophysicists concerned with water are called ____________.
15. ____________ study surface and underground waters in the landmasses of the earth.
16. Geophysicists work both ____________ and in the laboratory.
17. For an entry-level geophysical position, a ____________ degree is required.
18. Water covers over ____________ % of the planet's surface.
19. Careers in the ____________ sciences require specialized knowledge of the principles and techniques of the natural sciences, mathematics, and engineering.
20. Aquatic science is the study of our planet's ____________ waters.
21. ____________ is the study of the biological, chemical, geological, optical, and physical characteristics of oceans and estuaries.
22. ____________ involves the study of the characteristics of inland systems, including both fresh and salt waters.
23. ____________ do research in developing fisheries and mining the ocean in sustainable ways.
24. Some examples of research conducted by oceanographers and limnologists include:

    a.

    b.

    c.

    d.

    e.

    f.

    g.
25. Aquatic scientists spend part of their time working in offices and the remainder of the time in the ____________.
26. For oceanographers, ____________ means getting out onto the ocean: collecting, measuring, recording, and tagging fish, crustaceans, marine mammals, and other ocean creatures.
27. A minimum of a bachelor's degree is required for work at the professional level in ____________.
28. ____________ study the air that surrounds the earth.
29. The best-known application of meteorologists is weather ____________.
30. Trying to predict when it will rain and how much rain will fall is the job of the ____________.
31. A meteorologist who deals with water supplies, sunshine, and temperature is called an ____________ meteorologist.
32. ____________ meteorologists study the atmosphere's chemical and physical properties.
33. ____________ study climate variations over long periods of time.
34. ____________ meteorologists study and report on air quality.
35. Meteorologists are expected to work day, night, or on some type of ____________ plan consisting of a total of 40 hours a week.

36. The largest employer of meteorologists is the ____________ government.
37. ____________ water is absolutely essential for the smooth functioning of our society.
38. Water ____________ plant operators treat water so that it is safe for human consumption.
39. ____________ treatment plant operators treat sewage and other wastewater to remove pollutants and neutralize pathogens so that the water is safe to return to the hydrologic cycle.
40. As our population grows, the pressure to provide adequate supplies of clean potable water ____________.
41. With the passage of the Clean Water Act of 1972, water pollution standards have become ____________.
42. The Safe Drinking Water Act of 1974 established ____________ for water for human consumption.
43. Water or wastewater treatment plant operators usually must have at least a ____________ school diploma.
44. The Safe Drinking Water Amendment of 1996 specifies national minimum standards for ____________ and recertification of operators.
45. Employment of both water treatment plant operators and wastewater treatment plant operators is expected to ____________.

ACTIVITY

Purpose:

Determine the amount of water used.

Procedure:

Complete the following:

1. The average person will take a shower ______________ times per day.
2. Two showers per day × 30 gallons = ______________ gallons per day.
3. ______________ $\frac{\text{gallon}}{\text{day}}$ × 30 gallons per month = ______________ gallons per month
4. ______________ $\frac{\text{gallon}}{\text{month}}$ × 12 months per year = ______________ gallons per year.
5. ______________ $\frac{\text{gallon}}{\text{year}}$ ÷ 7.46 gallons per cubic foot = ______________.

Observations:

1. How many gallons per year are used by the average person when showering?
2. How many cubic feet per year are used by the average person when showering?

JOB EXERCISE FOR CHAPTERS 13–18

Purpose:

Set up a sprinkler system.

Materials Needed:

valve

elbow

adaptor

3/4-inch PVC pipe

tee

riser

head

PVC cement

PVC pipe cutter

Procedure:

1. Cut two 1-foot sections and one 2-foot section of *3/4-inch PVC pipe*, using the *PVC pipe cutter*.
2. Screw the *adaptor* into the valve outlet.
3. Install the *adaptor* to the end of a 1-foot section of *3/4-inch PVC pipe* and an *elbow* at the other end.
4. Connect the 2-foot section of *3/4-inch PVC pipe* to the *elbow*.
5. Connect the *tee* to the other end of the 2-foot section of pipe.
6. Connect the *riser* to the leg of the *tee*.
7. Screw the *head* to the *riser*.
8. Connect the other 1-foot section to the outlet of the *tee*, and connect the cap to the end of the *3/4-inch PVC pipe*.
9. Once the system is complete, check with the instructor before gluing with *PVC cement*.

Observations:

Compare your completed system to the following picture:

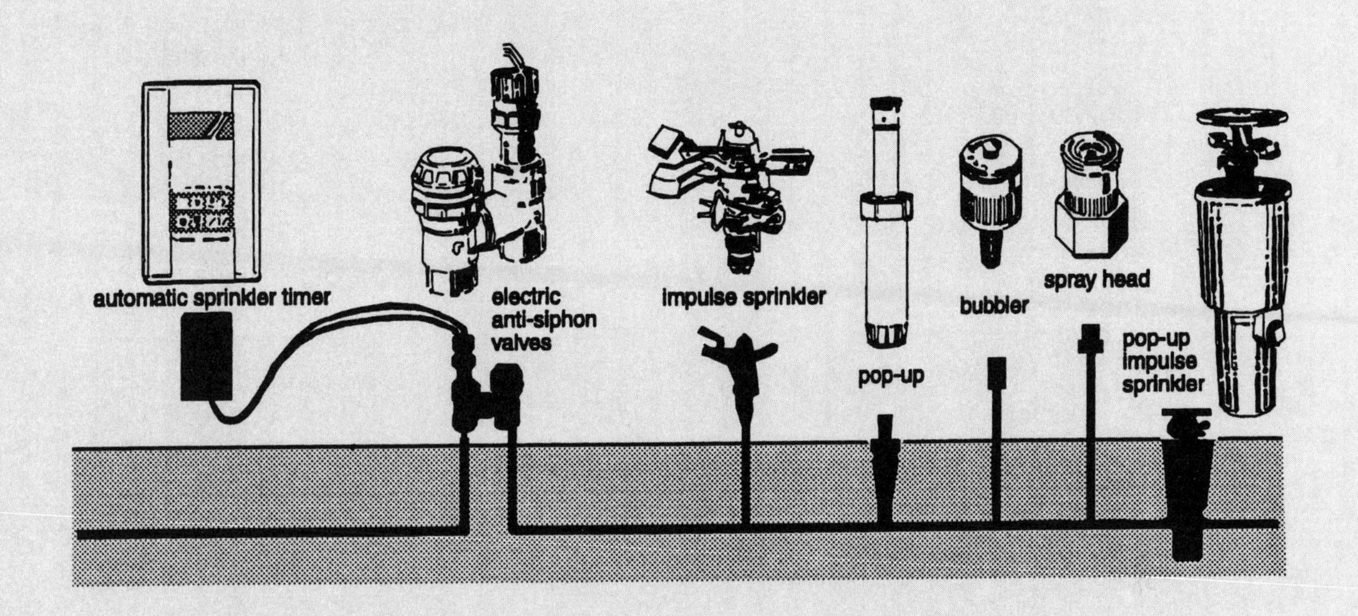

Forest Resources

CHAPTER

19

Our Forests and Their Products

TEST YOUR KNOWLEDGE

Complete the following:

1. Leif ____________ landed in North America about 1,000 years ago.
2. Erickson's workers harvested ____________ to send home by ship.
3. Our forests and their ____________ and by-products have always been important.
4. Forests are one of our ____________ resources.
5. Trees can be harvested and then ____________ by new trees.
6. A ____________ is a highly complex community of trees, shrubs, other plants, and animals.
7. Forests were both an ____________ and a friend to the settlers.
8. When the first permanent English settlement in America was established at ____________, in the Colony of Virginia, trees were everywhere.
9. The first task of the settlers was to provide protection and ____________ for themselves.
10. The need to establish a reliable food supply meant ____________ the forest.
11. Trees were ____________ but the wood could then be used to build forts, homes, and other structures.
12. The ____________ building industry of Europe came to depend on the North American colonies for tall, straight trees for their ships.
13. The ____________ stores industry grew up around harvesting tar and pitch from pine trees in the southern forest region.
14. According to the United States Department of Agriculture (USDA), the total land areas in what is now the United States is about ____________ billion acres.
15. Roughly one-third of U.S. forest and woodland is referred to as ____________ forestland.
16. About ____________ million acres of U.S. forest is generally referred to as commercial forest.
17. ____________ forests simply means that the land is capable of producing economically useful forest.
18. Of the 481 million commercial forest acres, ____________ individuals own 58%, or 279 million acres.

19. List the forest products industries.

    a.

    b.

    c.

20. The typical ____________ farm is a single-family operation of less than 200 acres in size.
21. In the United States today there are ____________ species of trees.
22. List the things that affect the kinds of tree that will grow in the United States.

    a.

    b.

    c.

23. The six major continental U.S. forest regions are:

    a.

    b.

    c.

    d.

    e.

    f.

24. List the four forest regions in Alaska and Hawaii.

    a.

    b.

    c.

    d.

25. The three main parts of a tree are:

    a.

    b.

    c.

26. The root system serves to ____________ and support the tree.
27. Other functions of the root system include:

    a.

    b.

    c.

    d.

    e.

28. List the four types of roots in a complete root system.

    a.

    b.

    c.

    d.

29. Roots grow both in ____________ and in diameter.
30. Growth in length is accomplished by an area of rapid ____________ division.

31. Root ____________ absorb the water and nutrients taken in by the root system.
32. The trunk ____________ the crown of the tree.
33. List the five parts of mature tree trunk.
    a.
    b.
    c.
    d.
    e.
34. The ____________ consists of woody cells that are inactive.
35. The ____________ is a thin layer of active cells that divides to produce new cells.
36. Inner bark, also known as ____________, consists of living cells by which the food produced in the leaves moves downward to the roots and to the rest of the trees.
37. In the ____________, new sapwood cells are large and soft walled, called springwood.
38. In the ____________, new cells are smaller and darker in color, called summerwood.
39. The ____________ of the tree can be determined by counting the annual rings.
40. During a year with an extremely good growing season, the annual ____________ will be very wide.
41. During dry years, or when disease, fires, insects, or other problems arise, the rings may be extremely ____________.
42. The crown of the tree includes:
    a.
    b.
    c.
    d.
43. What are the two ways that growth in the crown takes place?
    a.
    b.
44. Leaves convert water and carbon dioxide in the presence of sunlight into sugar in a process called ____________.
45. Excess water from the tree is allowed to evaporate through openings in the leaves in the process called ____________.
46. Trees can be classified as either shade ____________ or shade intolerant.
47. Shade ____________ trees grow satisfactorily without complete direct sunlight.
48. Trees that require some direct sunlight but can also grow in partial shade are known as ____________ shade tolerant.
49. Shade-intolerant trees do not fare well without ____________ sunlight for at least part of the day.
50. A ____________ forest has mostly a single species of tree.
51. ____________ aged forests have trees of a single age and size.
52. ____________ aged forests include trees in two or more size groupings.
53. The ceiling called the ____________ of a forest is made up of the crowns of the taller trees.
54. Members of the forest canopy include:
    a.
    b.
    c.
    d.

55. Lumber grades include:

    a.

    b.

    c.

    d.

    e.

    f.

56. Once the wood is mechanically or __________ changed, it is referred to as converted wood.
57. ______________ woods include products such as paper, pulp, wood fiber, charcoal, explosives, and plastics.
58. Other forest-produced benefits include:

    a.

    b.

    c.

    d.

59. Forests are a major contributor to the balancing of ______________ and carbon dioxide in the atmosphere.
60. No other covering, including concrete, is so ______________ of the soil in the long run.

ACTIVITY

Purpose:

To identify the parts of a tree.

Procedure:

1. Review the section titled "Trees and Their Growth" in the text.
2. Label the following pictures.

A.

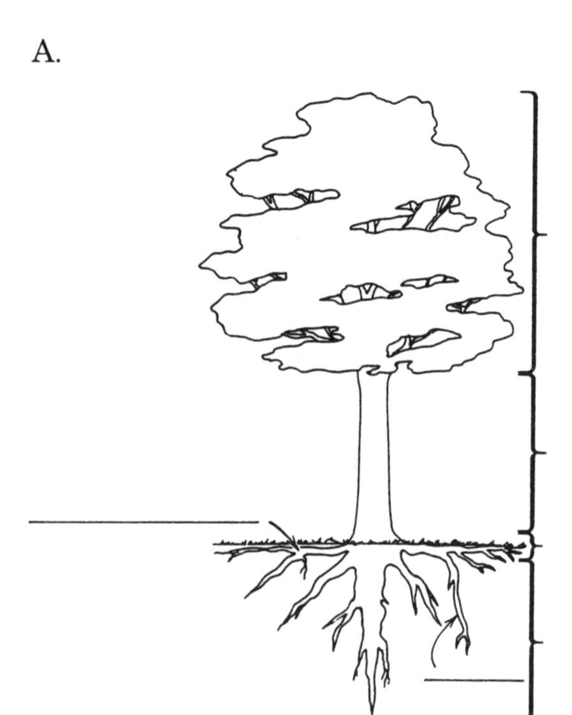

B.

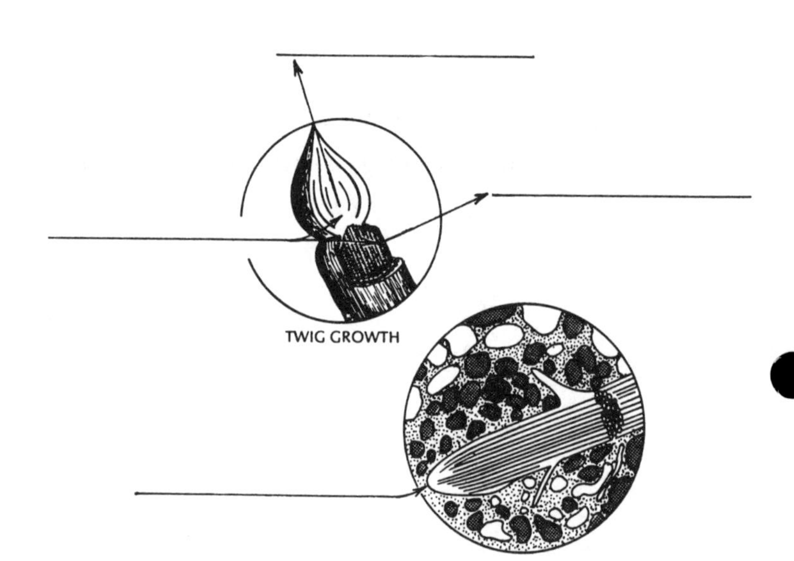

C.

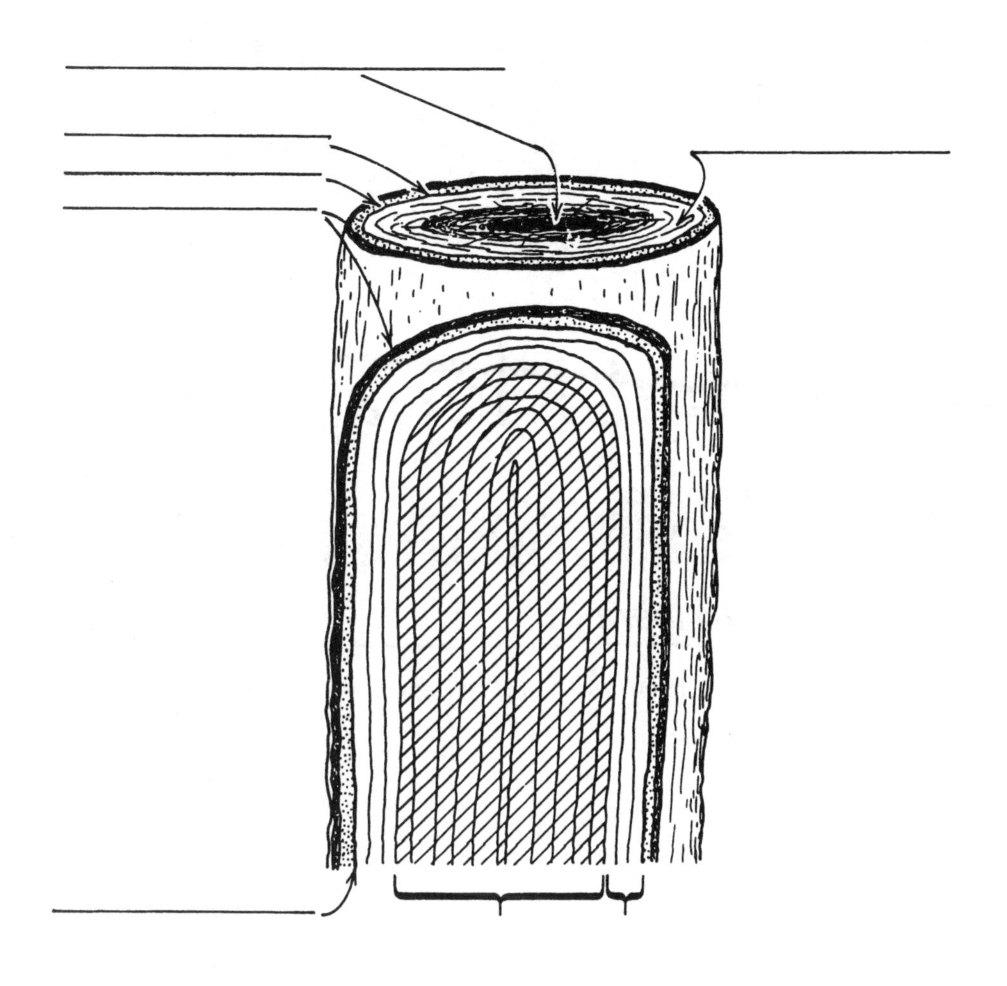

Observations:

1. Study the parts of the tree.
2. Study the parts of tree branches and roots.
3. Study the parts of the tree trunk.

CHAPTER
20

Woodland Management

TEST YOUR KNOWLEDGE

Complete the following:

1. List the ways that forest resources are of great importance, as mentioned in this chapter.
   a.
   b.
   c.
   d.
2. The most basic consideration in modern forest management is the sustainable production of ___________.
3. Foresters can ___________ the amount of wood before the trees are cut.
4. The standard unit of measure for most lumber is the ___________ foot.
5. Timber and lumber are usually sold at a ___________ per board foot.
6. A board foot is a piece of ___________ wood 1 foot long, 1 foot wide, and 1 inch thick.
7. A board foot need not be an exact shape, but the volume is always ___________ cubic inches.
8. The formula for ___________ feet is length times width times thickness, divided by 144.
9. A ___________ foot is the amount of wood that fills a space 1 foot wide, 1 foot thick, and 1 foot high.
10. ___________ is still often sold by the cord.
11. A standard ___________ of wood is a stack, 4 feet by 4 feet by 8 feet.
12. A standard cord of wood converts to ___________ cubic feet.
13. Gross weight is the most common method of measuring ___________.
14. The ___________ of wood in a log is determined by its diameter and length.
15. For a tree that is still standing, the diameter is measured "at ___________ height."
16. Diameter at breast height (DBH) is the thickness across the tree trunk at ___________ feet above the average ground level.
17. DBH is usually rounded to the nearest ___________ inch class.

18. The three most common dendrometers are:

 a.

 b.

 c.

19. Tree height is the ___________ length of the trunk.
20. For sawtimber, tree height is expressed in terms of ___________ foot logs.
21. A ___________ is the length to which each pulpwood log is cut.
22. Tree height is measured from the height of the ___________ to the point on the trunk where the cutoff diameter is estimated.
23. Tree height is measured by a ___________.
24. The most common hypsometers are:

 a.

 b.

 c.

 d.

25. Log ___________ or tables are constructed to estimate the volume of the tree.
26. The most common log rules are:

 a.

 b.

 c.

27. Estimating ___________ timber volume is known as cruising timber.
28. Cuttings taken during the time from planting or natural ___________ to harvest are called intermediate cuttings.
29. Intermediate cuttings to assist young seedlings or saplings are called ___________.
30. ___________ cuttings in older trees are called improvement cuttings.
31. The total timber volume produced by a stand can be increased by removing some of the trees, a procedure called ___________.
32. Removing dominant, taller trees is called ___________.
33. ___________ kills the tree but leaves it standing until it decays and falls on its own.
34. ___________ cutting removes injured, diseased, or insect-infested trees.
35. A ___________ cutting is done when the damaged trees are harvested and sold for wood.
36. ___________ burning is the controlled burning of the undergrowth in a pine forest of saplings to -mature trees.
37. In ___________ cutting, trees are selected based on maturity, size, species, growth rates, and other factors.
38. When a forest is harvested in two or three stages, it is called ___________ cutting.
39. In ___________ cutting basically the entire stand of trees is removed at harvest, and only a few of the best trees are left standing.
40. ___________ cuttings cut most of the trees at the initial harvest, but trees called standards are left.
41. Many hardwood trees will produce sprouts or suckers from their stumps or roots after they have been cut, and this is called ___________ growth.
42. In ___________ cutting, all the trees of a tract are harvested in a single operation.

43. The four methods of reproducing the forest include:

    a.

    b.

    c.

    d.

44. The quickest way to get a full stand of trees is to plant ____________-produced seedlings.
45. Commercial tree ____________ produce and sell large quantities and many varieties of trees for forestry use.
46. List the items discussed in this chapter for a good forest management plan.

    a.

    b.

    c.

    d.

ACTIVITY

Purpose:

To determine the volume of a log.

Research:

1. Volume tables are used to determine the volume of a standing tree or cut log.
2. Volume tables can show cubic feet or board feet for an area.
3. Tables have the height across the top, and DBH on the left side.

Procedure:

Review the table given on the next page and practice using it.

Observations:

Complete the following using a board feet volume table.

Tree	Height in feet	DBH in inches	Board feet
1	100	24	
2	80	16	
3	75	15	
4	95	20	
5	90	25	

Gross Volume per Tree for Ponderosa Pine in the Black Hills
Board Feet in Scribner Rule

Board Feet Inside Bark — Top Diameter—6.0 Inches

Merchantable Stem Excluding Stump and Top — Stump Height—1.0 Foot

Total Height

DBH	20	25	30	35	40	45	50	55	60	65	70	75	80	85	90	95	100	105	110	115	120	125
8	1	2	4	6	9	11	13	15	17	19	22	24	26									
9	3	6	8	11	14	16	19	22	24	27	30	33	35	38								
10	6	9	13	20	27	33	40	47	54	61	67	74	81	88	95							
11	8	12	21	29	37	45	53	61	69	78	86	94	102	110	118	127						
12	9	19	29	38	48	58	67	77	87	96	106	115	125	135	144	154	164					
13	15	27	38	49	60	72	83	94	105	117	128	139	150	161	173	188	203	218				
14	22	35	48	61	74	87	100	112	125	138	151	164	179	196	214	231	248	265	282			
15	29	44	58	73	88	103	118	133	147	162	179	199	218	238	257	277	297	316	336	355		
16	36	53	70	87	104	120	137	154	171	193	215	237	260	282	304	326	348	371	393	415	437	
17	44	63	82	101	120	139	158	179	204	229	254	279	304	329	354	379	404	429	454	479	504	529
18	53	74	95	116	138	159	183	211	239	267	295	323	350	378	406	434	462	490	518	546	574	602
19	62	86	109	133	156	183	214	245	276	307	338	369	400	431	462	493	524	555	586	617	648	679
20	72	98	124	150	178	212	246	281	315	349	383	418	452	486	521	555	589	623	658	692	726	761
21	82	111	139	167	205	243	280	318	356	394	431	469	507	544	582	620	658	695	733	771	808	846
22		124	155	192	234	275	316	357	399	440	481	523	564	605	647	688	729	770	812	853	894	936
23			173	218	263	309	354	399	444	489	534	579	624	669	714	759	804	849	894	939	984	1029
24				246	295	344	393	442	491	540	589	638	686	735	784	833	882	931	980	1029	1078	1127
25					327	380	433	486	539	593	646	699	752	805	858	911	964	1017	1070	1123	1176	1229
26						418	476	533	590	648	705	762	819	877	934	991	1049	1106	1163	1221	1278	1335
27							520	581	643	705	766	828	890	952	1013	1075	1137	1198	1260	1322	1384	1445
28								632	698	764	830	897	963	1029	1095	1162	1228	1294	1361	1427	1493	1559
29									755	826	897	968	1039	1110	1181	1252	1323	1394	1465	1536	1607	1678
30										889	965	1041	1117	1193	1269	1345	1420	1496	1572	1648	1724	1800

CHAPTER

21

Forest Enemies and Their Control

TEST YOUR KNOWLEDGE

Complete the following:

1. List the forest enemies.

   a.

   b.

   c.

   d.

   e.

   f.

2. Trees are ________________ to grow best in certain climates, and under certain conditions.
3. Forest enemies include anything that interferes with the normal reproduction and ________________ of forest trees.
4. According to some authorities, ________________ kill more trees than any other enemy of the forest.
5. The main insect types that cause damage include:

   a.

   b.

   c.

   d.

   e.

   f.

   g.

   h.

6. The bark beetle can ________________ completely around the tree, girdling it.
7. When a tree is ________________, it can no longer send food from the leaves to the roots.
8. ________________ insects damage trees by feeding on the leaves or needles.
9. ________________ insects eat their way through the sapwood and heartwood of the tree.
10. Insects that attack young twigs, stems, or buds are ________________ feeders.
11. Sap suckers usually only ________________ trees and slow their growth.
12. ________________ are abnormal growth and usually cause little serious damage.
13. ________________ eaters live by eating tree nuts, fruits, and seeds.
14. Insects in the ground may feed on tree ________________.
15. Insects have ________________ enemies.
16. Controls of insects include:

    a.

    b.

    c.

17. Whenever foresters alter natural controls, we say they are using ________________ insect controls.
18. A ________________ is any disease-causing organism.
19. Trees that have been damaged are susceptible to ________________ attack.
20. Damaged trees should be harvested or destroyed because:

    a.

    b.

    c.

21. Researchers have been working for many years to develop better varieties of trees, and one aspect they have worked on is ________________ resistance.
22. When trees are too crowded, they are more ________________ to insect and disease problems than healthy trees.
23. ________________ can provide a quick and effective short-term solution.
24. Forest ________________ is the study of tree diseases, their characteristics, causes, prevention, and treatment.
25. Noninfectious diseases are usually caused by ________________ problems.
26. Infectious diseases are usually caused by ________________.
27. ________________ are organisms that rely on other organisms for their food.
28. The five major groups of disease-causing organisms for trees are:

    a.

    b.

    c.

    d.

    e.

29. Fungi reproduce and spread by tiny ________________.
30. Fungus diseases attack the leaves, stems, or ________________of the tree.
31. The most effective control of forest diseases is good ________________.
32. Fomes rot can be prevented when the stumps are dusted with borax or ________________.
33. Branches of trees damaged by stem fungus should be pruned and ________________.

34. Damaged trees are ________________ subject to disease.
35. Crowded trees are more subject to disease, and forest stands should be ________________ as needed for best tree density.
36. ________________ -resistant varieties have been (and are being) developed for many species of commercial trees.
37. ________________ to kill disease carriers is often effective in preventing its spread.
38. When wildlife populations ________________, damage to the forest can result.
39. The best solution to dense wildlife populations is a well-managed ________________ and trapping program.
40. Overgrazing can cause ________________ damage to the forest.
41. Sheet ice storms and heavy ________________ can break limbs and even whole trees.
42. Well-managed forestry ________________ the amount of damage the forest will suffer.

ACTIVITY

Purpose:

To identify forest insects and their damage.

Research:

1. Review the following list of insects:

   Aphid
   Scale
   Budworm
   Spider mite
   Beetle
   Borer
   Sawfly

2. Go to your local library or use the Internet to research information on the insects listed in item 1.
3. Review the information about these insects.

Procedure:

Complete the following table:

Insect	Identification	Insect damage
Aphid		
Scale		
Budworm		
Spider mite		
Beetle		
Borer		
Sawfly		

Observations:

1. Which of the insects cause damage to pine trees?
2. Which of the insects cause damage that is not visible on the exterior of the tree?

CHAPTER 22

Fire!

TEST YOUR KNOWLEDGE

Complete the following:

1. The original ____________ Bear was actually a survivor of a forest fire.
2. Native Americans used fire as a way to clear the land and to improve ____________.
3. Managed, intentional fire is called a ____________ fire.
4. Benefits of a prescribed fire include:
   a.
   b.
   c.
   d.
   e.
   f.
   g.
   h.
5. A prescribed fire must be ____________ very closely.
6. The prescribed fire must not be allowed to get too ____________.
7. The prescribed fire must not be allowed to get out of ____________.
8. The most famous fire in U.S. history is the Great ____________ Fire.
9. The great ____________ fire damaged 1.3 million forest acres in Wisconsin, and 2.5 million forest acres in Michigan.
10. In general, the forest fire ____________ programs in the United States have been hailed as one of the most successful environmental programs in the country's history.

11. A fully mature forest may be beautiful, but it is very ____________ from many perspectives.
12. A forest can remain healthy and growing if periodically is:
    a.
    b.
    c.
13. ____________ are part of the natural system.
14. National efforts at preventing and controlling forest fires have resulted in unnaturally heavy production of ____________ and trees weak from over-competition.
15. There is a ____________ danger of major forest fires, forest disease, and insect infestations.
16. We might be wise to ____________ our future attempts at preventing forest fires.
17. Fire is both a chemical and a ____________ process.
18. Forest fires result from the rapid combination of ____________ with other chemicals contained in wood, leaves, and other material in the forest.
19. For a forest fire to occur, three things are required:
    a.
    b.
    c.
20. To ____________, a forest fire must have all three sides of the fire triangle.
21. Most fires are caused by ____________.
22. Major sources of wildfire include:
    a.
    b.
    c.
    d.
    e.
    f.
    g.
23. The three categories of forest fires include:
    a.
    b.
    c.
24. Once a surface fire gets into the forest canopy, it becomes a ____________ fire.
25. An ____________ is used to determine the direction from the tower to the fire.
26. The forestry office uses a process called ____________ to determine the actual location of the fire.

27. Factors that affect the anatomy of a fire include:
    a.
    b.
    c.
    d.
    e.
    f.
    g.
    h.
    i.
28. The best way to combat fires is to ____________ them from starting in the first place.
29. The National Fire Danger Rating Service predicts fire potential using:
    a.
    b.
    c.
30. When the burning index is particularly high, extra ____________ is needed.
31. The best method of fire suppression is ____________ attack.
32. An ____________ attack on a wildfire involves removing fuel from the fire triangle.
33. A fire ____________ may be any road, stream, bare area, field, lake, or other natural obstacle to the fire.
34. ____________ are normally constructed or completed first in front of the head of the fire because the fire spreads most quickly in that direction.
35. ____________ are smaller fires set along the firebreak on the side toward the fire.
36. Ground crews must constantly ____________ the fire line.
37. ____________ up involves patrolling of the fire line until the fire is no longer dangerous.

ACTIVITY

Purpose:

To identify forest tools and equipment.

Research:

1. Locate a forestry suppliers catalog.
2. Find each tool in the catalog index.

Procedure:

Complete the following table.

Tool/equipment	Item number	Price
Tree stick		
Diameter tape		
Increment borer		
Bark gauge		
Tree calipers		
Forester axe		
Hand compass		
Plastic flagging		
Clinometer		
Safety hat		

Observations:

1. Which of the listed tools is the most expensive?
2. Which of the listed tools have you not used?

CHAPTER 23

Careers in Forestry

TEST YOUR KNOWLEDGE

Complete the following:

1. The forest and forest products industries in this country make up a ____________ dollar part of our economy.
2. The forest industry produces about ____________ % of the gross national product in the United States.
3. Almost ____________ % of all jobs in the United States are in the forestry and forest products industries.
4. List the areas of from which over 3 million jobs of forestry come from.
   a.
   b.
   c.
   d.
   e.
5. The occupation of____________ is the professional who helps manage, direct, and protect our nation's forest resource.
6. List five responsibilities of a forester.
   a.
   b.
   c.
   d.
   e.
7. List the technology tools used by foresters.
   a.
   b.
   c.
   d.
   e.

8. Range managers and soil ____________ are the two most common forestry scientists.
9. Range managers study, manage, improve, and ____________ rangelands.
10. In May 2012, there were ____________ foresters in the United States.
11. Employment opportunities for foresters are expected to ____________ from 2010 to 2020.
12. To become a forester you need at least a ____________ education.
13. People who work as forestry technicians are also known as:
    a.
    b.
    c.
14. The work of the forestry technicians is typically done under the ____________ of a professional forester or other conservationist.
15. Name five things that forestry technicians are responsible for.
    a.
    b.
    c.
    d.
    e.
16. It is predicted that there will be ____________ in this occupation between 2010 and 2020.
17. Most entry level forest and conservation workers learn from experienced workers through ____________.
18. The first task of a logger is ____________ the trees.
19. A tree is ____________ in such a way that its fall does the least damage.
20. An ____________ removes a wedge of wood from the side where the logger wants the tree to fall.
21. The ____________ is made just above the undercut on the opposite side.
22. When a tree is cut into logs or bolts it is called ____________.
23. ____________ is when the logs are dragged to a loading area where they are sorted by size and mechanically loaded onto trucks for hauling.
24. Jobs in logging are expected to ____________ than average through 2020 because of mechanization in the logging industry.
25. Every time machine replaces muscle, the requirement for training ____________.
26. List the three levels at which forestry subjects are taught.
    a.
    b.
    c.

ACTIVITY

Purpose:

Identify the responsibilities of a forester.

Research:

1. Choose a forestry career.
2. Using various career sources, research and locate information about this career.

Procedure:

List the sources used and the information obtained from these sources.

Observations:

1. What is the nature of work for this career?
2. What is the employment outlook for this career?
3. What training is required for this career?

JOB EXERCISE FOR CHAPTERS 19–23

Purpose:

Measure the diameter of a tree.

Materials Needed:

Diameter tape
Carpenter's tape measure
Safety equipment as needed (gloves, hard hats, eye and ear protection)

Procedure:

1. Assemble the needed tools and equipment and carry them to the site where the trees are to be measured.
2. Using the carpenter's tape, measure to a height of 4.5 feet above ground level on the uphill side of the tree.
3. Secure the end of the diameter tape to the bark at the 4.5-foot height level.
4. Extend the diameter tape around the tree until it meets the end.
5. Read the diameter directly from the tape; tree diameters are usually recorded in the nearest 2-inch class (8 inch, 10 inch, 12 inch).
6. Return the tools and equipment to the storage area, and put them away properly.

Observations:

Complete the following table for the trees you measured.

Tree	Diameter
1	
2	
3	
4	
5	

Fish and Wildlife Resources

CHAPTER

24

Fish and Wildlife in America

TEST YOUR KNOWLEDGE

Complete the following:

1. _____________ of our renewable resources, such as birds, mammals, and fish, is just as important as management of our coal and oil.
2. A common definition would define wildlife as living things that are neither human nor domesticated, especially _____________, birds, and fishes.
3. The continental United States contains over _____________ different vertebrate species.
4. Without _____________ resources, the wilderness would have never been conquered.
5. The United States was established in the world _____________ trade business at the expense of the wildlife.
6. Many wildlife species were killed because they appeared _____________.
7. To survive, any species, plant, or animal must _____________.
8. Every species of plant or animal requires an area in which it can find shelter, food, water, and safety, called _____________.
9. In any habitat for any given species of wildlife, there is always a _____________ factor on the population of that species.
10. List the three of the limiting factors on a population.

    a.

    b.

    c.
11. The population limit is known as the _____________ capacity for that species in that habitat.
12. The _____________ curve or the population curve suggests that the population of the organism will increase slowly but then will start to increase very rapidly.
13. In the absence of significant numbers of _____________, the deer population in North America has increased steadily over recent decades.
14. In every state, deer hunting is managed by the _____________ fish and wildlife service.

15. Deer seasons and bag limits are set to help maintain a ____________ deer population.
16. Many species of wildlife have become so few in numbers they are considered rare, threatened, or ____________.
17. A species of animal is considered ____________ only when limited numbers of breeding pairs are known to exist, but is not in immediate danger of becoming extinct.
18. A(n) ____________ animal is one that is in danger of becoming extinct in all or major parts of its habitat.
19. A ____________ species is one that is likely to become endangered in the foreseeable future.
20. The Endangered Species Conservation Act was enacted to ____________ fish and wildlife on a worldwide basis.
21. The Threatened and Endangered Species System (TESS) Box ____________ summarizes the status of all species listed as threatened and endangered worldwide.
22. ____________ species no longer exist outside of museums and photographs.
23. A ____________ or endangered species is one that is no longer common and is in danger of becoming extinct.
24. The last-known ____________ pigeon died in 1914 in a Cincinnati zoo.
25. The ____________ parakeet's feathers were prized for decorating women's hats.
26. A ____________ swept across the sanctuary of the heath hen.
27. Using their feathers to stuff ____________ caused the extinction of the Labrador duck.
28. There are 363 mammals worldwide on the endangered list distributed by the U.S. Department of the ____________.
29. The Bighorn sheep are threatened by extinction from ____________ and disease.
30. In addition to hunting, loss of ____________ is causing significant problems for the polar bear.
31. The smallest white tail deer is the ____________ deer, once sought for trophies; now laws prohibit the hunting of the deer.
32. There have been bounties on the ____________ since ancient Greece when a reward was offered to any hunter who killed one.
33. Pumas or mountain lions are hunted with ____________.
34. The U.S. Department of the Interior has placed ____________ birds on the endangered species list.
35. The whooping cranes fly from Texas to Canada each year on their ____________, and many hunters have shot these large white targets.
36. According to the 2013–14 survey, biologists estimate a total population of more than ____________ whooping cranes remaining.
37. The bald eagles were killed mainly because it was believed they fed on ____________.
38. The ivory-billed woodpecker, if not ____________, is one of the most endangered species of bird known today.
39. The Prairie ____________ Foundation was formed to work for the management of the prairie chicken.
40. As of December 5, 2013, there were ____________ fish on the Department of Interior endangered species list.

ACTIVITY

Purpose:

Compile a profile of an endangered species.

Research:

Choose an endangered species you are interested in. Research this species by going to your local library and keeping notes.

Procedure:

1. Compile a one-page report about your chosen rare or endangered species.
2. Include the following information:

   Description
   Habitat
   Reproduction
   Role of animal
   Future
   Summary

Observations:

1. What habitat is the most critical for this animal?
2. What would you do to improve the habitat for this animal?

CHAPTER 25

Game Management

TEST YOUR KNOWLEDGE

Complete the following:

1. When European ____________ first came to the United States, they were greeted with a vast array of plant and animal life.
2. Aldo ____________, an early authority on game management, referred to game management as the "art of making land produce sustained annual crops of wild game for recreational use."
3. Scientists define game management as the science and art of changing characteristics and ____________ of habitats, wild animal populations, and humans to achieve specific human goals by means of the wildlife resource.
4. List the five basic habitat requirements, as listed in this chapter.

   a.

   b.

   c.

   d.

   e.
5. Game animals can be classified by both the type and amount of ____________ they consume.
6. List the six listed classifications of game animals that deal with the type of food they consume.

   a.

   b.

   c.

   d.

   e.

   f.
7. ____________ are plant eaters.
8. ____________ are meat eaters.

9. ____________ are insect eaters.
10. ____________ are fruit eaters.
11. ____________ are animals that eat many different types of food.
12. ____________ are seedeaters.
13. The classification of wildlife by food quantity includes:
    a.
    b.
14. A ____________ animal is one that consumes great varieties of food.
15. Because its choice of food is usually varied, a euryphagous' chances of survival are usually ____________.
16. A ____________ animal is one that eats a specialized diet.
17. A stenophagous animal has ____________ chance to adapt to new food sources if its traditional food supply is not available.
18. For game to survive harsh weather conditions, they must find ____________, a place that will protect them.
19. ____________ is one of the most important requirements of wildlife.
20. The bodies of most game animals consist of 60 to ____________ % water.
21. The area over which the game travels is called its home ____________.
22. The area an animal will defend is called its ____________.
23. Many animals' home ranges can overlap, but their ____________ almost never will.
24. The most commonly accepted game management procedures include:
    a.
    b.
    c.
    d.
    e.
    f.
25. List the game management procedure that set aside land for the protection of wildlife.
    a.
    b.
    c.
26. The first state reserve was established in 1870 in ____________.
27. Refuges protect the wildlife only from hunters, not from their ____________ enemies.
28. To increase game populations, the development and improvement of their ____________ must be an integral part of game management procedures.
29. The most common habitat developments are ____________ plantings and woodland management.
30. The fencerow area once used by game for cover and food has become ____________.
31. The management of game as a resource is not ____________ of other resources.
32. In 1646, Rhode Island became the first state to establish a ____________ season on game.
33. The first bag limit was initiated in 1878 in the state of ____________.

34. List the factors that determine game populations.

a.

b.

c.

d.

e.

35. One facet of game management has been to control the ____________.

36. Predators can be ____________ to the game management plan.

37. By controlling the predator's population we can also control the population of the ____________ species.

38. Predators feed on animals that humans consider ____________, such as rats and mice.

39. Predator control practices keep the target species in a ____________ condition.

40. Predators help maintain an improved game population by killing the ____________ and injured individuals.

41. ____________ stocking is either the stocking of game natural to the area or the introduction of species new to the area.

42. Bringing in new species is called the "introduction of ____________."

43. Whenever humans try to manipulate game artificially, what two important principles must be carefully examined?

a.

b.

44. Population ____________ refers to the number of game animals in a defined area.

45. ____________ capacity refers to the amount of game an area will provide for the essentials of life.

46. The farm pond is a practical way to supply water for fish production and game animals, as well as for livestock and ____________ protection.

47. If the game animals do not have adequate ____________ to protect them from the elements, their chance for survival will be slim.

48. Government agencies such as the U.S. Forest Service under the direction of the U.S. Department of Agriculture can help individuals or groups with their ____________ management program.

49. List the five areas that the Natural Resources Conservation Service can help a landowner.

a.

b.

c.

d.

e.

50. As early as 1896, the Supreme Court gave the management of wildlife to the ____________.

51. Name the Federal Legislation of game management listed in the chapter.

a.

b.

c.

d.

e.

f.

g.

ACTIVITY

Purpose:

Determine carrying capacity.

Research:

Population density refers to the number of game animals in a defined area. Carrying capacity refers to the amount of game for which an area will provide the essentials of life. A population is the number of some species living in a particular area. Carrying capacity of an area is influenced by many factors that cause a change in the habitat. As a result of these changes, the population levels rise and fall.

Procedure:

Graph the following comparison information of black-tailed deer.

Month	Good range	Poor range
December	62	26
May	50	25
June	90	35
July	84	31
December	63	27

100												
90												
80												
70												
60												
50												
40												
30												
20												
10												
0	Dec	Jan	Feb	Mar	Apr	May	June	July	Aug	Sept	Oct	Nov	Dec

Observations:

Indicate whether the following statements are true or false, by placing a T or F next to each statement.

_____________ 1. The population is higher in the spring/summer.

_____________ 2. The population is lower in the winter.

_____________ 3. The average population level on good range is higher than on poor range.

_____________ 4. Population levels continue to rise throughout the year.

CHAPTER

26

Marine Fisheries Management

TEST YOUR KNOWLEDGE

Complete the following:

1. The ____________ can be considered the last frontier on this planet.
2. ____________ represents scientist's efforts to classify the oceans based on one or more of those characteristics to help us explain how the oceans work.
3. List the methods to classify ocean zones.

   a.

   b.

   c.

   d.

   e.
4. The five ocean depth zones listed in the chapter are:

   a.

   b.

   c.

   d.

   e.
5. The supratidal and intertidal areas are ____________ the water level.
6. The ____________ zone starts at the water line at low tide.
7. The ____________ zone contains the continental slope and rise.
8. The ____________ zone is considered the ocean deep zone.
9. List the three zones of the ocean based on light penetration.

   a.

   b.

   c.

10. The ____________ zone is the part of the ocean where sunlight penetrates the water.
11. Two water movements of importance to this book are:
    a.
    b.
12. When the sun and the moon line up with the earth, a strong tide is produced. This exceptionally high tide is called a ____________.
13. When the sun and the moon are at right angles with each other an exceptionally low tide is produced, called a ____________.
14. List the four major groups that marine animal life can be divided into.
    a.
    b.
    c.
    d.
15. The most common microscopic animals are the ____________.
16. Zooplankton are the ____________ food for many species of fish.
17. Some zooplankton, called ____________, are zooplankton all of their lives.
18. Zooplankton populations are also composed of the larvae of commercially and recreationally important crustacean species referred to as ____________.
19. The seven main species of salmon are:
    a.
    b.
    c.
    d.
    e.
    f.
    g.
20. Salmon begin life in freshwater, but migrate to the ____________ to live and grow.
21. List the management techniques used to preserve and protect the salmon population, described in the chapter.
    a.
    b.
    c.
    d.
    e.
    f.
22. List the four ways that salmon is sold, as discussed in the chapter.
    a.
    b.
    c.
    d.
23. Tuna is a member of the ____________ family and the leading game fish in the United States.

24. The four most commercially important tuna, as mentioned in the chapter are:

 a.

 b.

 c.

 d.

25. The mature size of the northern bluefin is ____________ pounds.

26. Three methods used to catch tuna are:

 a.

 b.

 c.

27. A major problem with the netting procedure is the accidental, but unlawful, netting of the ____________ swimming above the tuna.

28. The most economically important marine shellfish are:

 a.

 b.

 c.

 d.

29. Wild shrimp can spawn a couple of times a ____________.

30. The shrimp are captured by shrimp ____________ towing nets in the ocean near shore to a few miles offshore

31. Oysters reach their reproductive prime in ____________ years and some oysters have lived up to 20 years.

32. There are more than ____________ families of crabs

33. The largest crab is the ____________ crab

34. The main mammals of the ocean include:

 a.

 b.

 c.

 d.

35. ____________ whales obtain their food by straining plankton.

36. The Whaling Commission has stopped the hunting of the following whales:

 a.

 b.

 c.

 d.

 e.

37. Seals are divided into three groups:

 a.

 b.

 c.

38. ____________ are the only tusked seals.

39. The area where a freshwater source opens into the ocean is called an ____________.

40. To artificially propagate marine animals in a man-made, controlled environment is called ____________.

41. Aquatic farming practices performed in freshwater and saltwater accounts for ____________ percent of all seafood consumed globally
42. Aquaculture practiced in the ocean is often referred to as ____________.
43. Fish cannot be claimed as ____________ as land wildlife can.
44. We once thought of the world's oceans and the ocean's harvest as ____________.
45. List the species of aquaculture production.
    a.
    b.
    c.
    d.
    e.
    f.
    g.

ACTIVITY

Purpose:

Calculate the dressing percent of fish.

Research:

Catfish usually yield 55–60% of their live weight in dressed form. The head, viscera, and skin are equal to about 40–45% of the weight of a whole catfish. The ideal size of a catfish for the retail market is a live weight of 1 lb. – 1.25 lbs. Dressed fish will weigh from 8 to 10 ounces dressed weight.

Example:

Live weight = 10 ounces

Example:

Dressing percent 60%

Live weight × Dressing percent = Dressed weight

$10 \times 0.60 = 6$ ounces

Procedure:

Complete the following table concerning dressing percent.

Live weight	Dressing percent	Dressed weight
16 oz.	60	
18 oz.	60	
20 oz.	60	
16.5 oz.	55	
24.5 oz.	55	

Observations:

1. What are some factors that might affect the dressing percent?
2. What are some ways that the dressing percent can be increased??

Freshwater Fishery Management

TEST YOUR KNOWLEDGE

Complete the following:

1. The American Sportsfishing Association estimates that recreational fishing in the United States has created nearly ____________ million jobs
2. Name the three zones a lake can be divided into.
   a.
   b.
   c.
3. The littoral zone is the ____________ zone and contains rooted vegetation.
4. One of the most important organisms in the littoral zone is the tiny greenish-brown ____________.
5. In the ____________ zone rooted vegetation is no longer present.
6. ____________ reaches to the bottom of the limnetic zone.
7. The bottom zone of a lake is the ____________ zone.
8. ____________ are the most common organism in the profundal zone.
9. List the main uses of a farm pond, as mentioned in the chapter.
   a.
   b.
   c.
   d.
   e.
   f.

10. The United States Department of Agriculture (USDA) indicates that the ideal family farm pond is less than ____________ acres.
11. An acre is any rectangular surface area of ____________ square feet.
12. List the four components of new pond management.

    a.

    b.

    c.

    d.
13. List the two kinds of artificial ponds.

    a.

    b.
14. ____________ ponds are impounded behind an earth embankment or dam.
15. ____________ ponds are made by digging a pit below the surrounding ground level.
16. Pond selection site considerations to remember are:

    a.

    b.

    c.

    d.
17. The ____________ is the land that drains into the pond.
18. Do not plant ____________ or shrubs on dams because the roots can weaken the structure.
19. The ____________ assures that water will never flow over the dam.
20. The pond basin should be planted with a ____________ crop.
21. A farm pond should be ____________ with both the right kinds as well as the right number of fish for its size.
22. Some common freshwater fish are:

    a.

    b.

    c.

    d.

    e.

    f.
23. The largemouth bass can be recognized by its large ____________ and horizontal dark stripe or blotches down its side.
24. The diet of the bass consists of aquatic ____________ and fish.
25. Adult bass feed day and night, but are primarily ____________ feeders.
26. The young bluegill eats tiny ____________ in the water.
27. As they mature bluegill feed on aquatic ____________, snails, small crayfish, and small fish.
28. Channel ____________ are usually stocked in a farm pond in combination with largemouth bass and bluegill.

29. The main management procedures for fisheries include:

    a.

    b.

    c.

    d.

    e.

    f.

30. ____________ is important in providing food, shelter, oxygen, and spawning and nesting habitats.
31. An overabundance of aquatic ____________ growth can create problems.
32. Some blue-green algae produce ____________ that are harmful to humans and fish.
33. Washout areas created by ____________ provide substrate for aquatic plants.
34. ____________ control involves using some other living organism, either plant or animal, to control aquatic plants.
35. The application of aquatic ____________ to control pond weeds is another pond management tool for the pond owner.
36. Fish ____________ is a management tool which can be employed in small ponds and large lakes, so that managers will know species diversity, abundance, and fish size.
37. Fish sampling methods are:

    a.

    b.

    c.

    d.

38. ____________ nets are fine mesh nets that entangle fish.
39. ____________ nets are used in a current or around a school of fish.
40. ____________ are nets held on the bottom by weights and on the surface by floats.
41. Spot ____________ involves poisoning a small area of a larger body of water.
42. Using a boat ____________ as a means of fish sampling involves applying alternating current into the water to stun the fish.
43. Angling fish with a fish hook and line is an effective method of fish sampling for members of the ____________ family.
44. Most lakes and ponds are equipped with a ____________ structure, which is necessary to completely drain the lake or pond.
45. Fish populations are removed by poisoning the lake with ____________ or sodium sulfite.
46. The three population adjustment procedures listed in the chapter include:

    a.

    b.

    c.

47. Fertilizing ponds can cause an increase in the production of ____________, a principal food source for fish.
48. The principal nutrients applied as fertilizer to a pond include:

    a.

    b.

    c.

49. In the winter, the bacteria decomposing the waste use most of the oxygen, thus increasing the incidence of ____________.
50. Most states have enacted ____________ controlling the fish taken from their public waters.
51. The main factors associated with water quality include:
    a.
    b.
    c.
    d.
52. Fish grow best if the temperature is above ____________ degrees F.
53. Oxygen deficiencies commonly occur as a result of bacterial ____________.
54. The ideal pH for fish appears to be between ____________ and 9.0.
55. ____________ can be caused by animal activity such wading cattle, bottom fish activity, and erosion.

ACTIVITY

Purpose:

Identify the characteristics of fish.

Research:

The most common stocking fish include bass, bluegill, and catfish. Not all fish are suited for pond stocking. Before a decision is made concerning which species to stock, their biology should be understood.

Procedure:

Review the information about common freshwater fish in the text.

Observations:

Distinguish between the characteristics of the common stocking fish by placing one of the following symbols next to the correct statement:

B — bass

Bl — bluegill

C — catfish

__________ 1. Have a large mouth

__________ 2. Feed on plants, insect larvae, frogs, and small bluegill

__________ 3. Average life span is 6–8 years

__________ 4. Spawn from May to August

__________ 5. Record is 22 pounds and 4 ounces

__________ 6. Dark stripe or blotches on side

__________ 7. Take about 3 years to reach a length of 12 inches

__________ 8. A food source for bass

__________ 9. Take three growing seasons to mature

__________ 10. Eat aquatic insects

__________ 11. Spawn readily in shallow water

__________ 12. Eat plankton

CHAPTER

28

Careers in Fish and Wildlife

TEST YOUR KNOWLEDGE

Complete the following:

1. Federal and state governments have an __________ to manage the natural resources of the country.
2. Careers in fish and wildlife management fall into two large areas:

   a.

   b.
3. Employment in fish and wildlife management is mostly by:

   a.

   b.

   c.

   d.

   e.

   f.
4. The governmental watchdog of the game management program is the conservation officer or __________ warden.
5. The conservation officer is responsible for __________ and upholding all federal and state statues, with particular attention to those related to the conservation of natural resources.
6. The conservation officer must possess knowledge of the following:

   a.

   b.

   c.

   d.

   e.

   f.

7. List the equipment operated by the conservation officer.
   a.
   b.
   c.
   d.
   e.
   f.
8. The conservation officer must be able to use:
   a.
   b.
   c.
   d.
   e.
   f.
9. The conservation officer's duties include:
   a.
   b.
   c.
   d.
   e.
   f.
   g.
   h.
   i.
   j.
   k.
10. The wildlife biologist deals with the __________ practices of game rather than the unlawful activities against game.
11. One important duty of the wildlife biologist is the planting and tending of __________ plots.
12. The wildlife biologist keeps records of the game __________ in the area.
13. In addition to food plots and hunter check-in stations, wildlife biologists are responsible for:
    a.
    b.
    c.
    d.
    e.
    f.
    g.
    h.
    i.

14. Resource managers must like the __________ since they must work in all extremes of weather.
15. A career as a fish and wildlife technician is for persons who like working with __________.
16. A fish and wildlife technician looks at the causes of __________ and pollution and works to establish hunting and fishing regulations.
17. A __________ manager is concerned with those species of wildlife that are considered endangered or rare.
18. The wildlife __________ may assist the game biologist.
19. The wildlife technician __________ the wildlife biologist's management programs to propagate wildlife species.
20. An animal __________ biologist controls rodents and other pests in wildlife settings.
21. A fish technician can assist __________ biologists, who work with habitats, spawning, and artificially grown young.
22. A fish hatchery manager supervises fish __________ operations.
23. A career as a fish and wildlife __________ requires a person who likes the outdoors.
24. The fish and wildlife technician is involved with __________ relations, frequently talking to sports enthusiasts, school officials, and individual citizens.
25. A fish and wildlife technician must possess good writing skills to prepare __________.
26. Even with increased leisure time activity, the employment opportunities offered by the federal and state governments are not expected to __________.
27. The main job of the fish __________ technician is to produce fish for food or game.
28. With the increased pollution of our rivers and streams, fish numbers have greatly decreased and people interested in __________ propagating fish are more and more in demand.
29. The three main areas of fish culture are:
    a.
    b.
    c.
30. The employee of a fish hatchery works with __________ new fish.
31. A fish wildlife conservation __________ assists in the fish conservation programs.
32. An experimental __________ technician assists biologists in the field and laboratory.
33. The basic requirement of the fish culture technician is an __________ degree in fish and wildlife technology or fishery management.
34. The fish culture technician may advance to a __________ manager or fisheries management specialist.
35. Damage to many marine areas by pollution and urban/industrial growth has caused the field of fisheries management to __________.
36. A __________ studies the structure, physiology, development, and classification of many kinds of animals both domestic and wild.
37. List the titles used to identify zoologists by the animal group they study.
    a.
    b.
    c.
    d.
38. A __________ is a collection of live, wild animals and zookeepers handle the feeding and diet of these animals.
39. A __________ of a museum of natural history collects and exhibits animals.
40. According to the Department of Labor statistics, the employment for zoologists will have __________ than average growth from 2010 to 2020.

ACTIVITY

Purpose:

Take a pH test.

Materials Needed:

A pH test kit
Water samples

Research:

The ideal pH for fish is between 6.5 and 9.0. The pH is the measure of relative amounts of acidity and alkalinity. A pH of 7.0 is neutral. A number less than 7.0 indicates that the water is acidic; a number greater than 7.0 indicates that the water is alkaline.

Procedure:

1. Locate a pH test kit.
2. Follow the instructions in the kit to make pH tests on the samples provided by the instructor.
3. Complete the following table:

Sample	pH
1	
2	
3	
4	
5	

Observations:

1. Which of the samples would be acceptable for fish production?
2. How might you change a pH condition that is not acceptable?

JOB EXERCISE FOR CHAPTERS 24–28

Purpose:

Complete an application for employment.

Procedure:

1. Always be prepared to fill out an application.
2. Look over the entire application before beginning to write.
3. Follow the directions on the form.
4. Write or print neatly.
5. Correct mistakes with a simple line through the word.
6. Be honest.
7. Answer all questions.
8. Make a copy of the application whenever possible.

Observations:

Complete the following application.

APPLICATION FOR EMPLOYMENT

Prospective employees will receive consideration without discrimination because of race, religion, color, sex, age, national origin or physical handicap.

Last Name First Middle

Date

Street Address

Home Phone
( )-

City, State, Zip

Business Phone
( )-

Are you over the age of 18?
☐ Yes ☐ No

If not, do you have valid working papers?
☐ Yes ☐ No

Position Desired:

Do you have any impairments – physical, mental or medical – which would interfere with your ability to perform the job for which you have applied?
☐ Yes If yes, please explain.
☐ No

Date Available for Work

Have you ever been convicted of a crime?
☐ Yes ☐ No

Charge, Date & Disposition:

Salary Expected

Are you legally eligible for employment in the U.S.? (Proof of citizenship or immigration status may be required upon employment.)
☐ Yes ☐ No

Have you ever been bonded?
☐ Yes ☐ No

How did you learn of our organization?

If the job requires, are you available to work overtime?
☐ Yes ☐ No

Indicate skills you possess:

Typing Speed _____ wpm Steno Speed _____ wpm

Computers (specify):

Software (specify):

List any professional memberships which you feel would be an asset for the position applied:

Military

Served in U.S. Armed Forces
☐ Yes ☐ No

Date Inducted:

Date Discharged:

Rank at Discharge

Briefly describe your duties:

EDUCATION

	High School	College/University	Graduate/Professional	Other
School Name				
Years Completed: (circle)	9 10 11 12	1 2 3 4	1 2 3 4	1 2 3 4
Diploma/Degree				
Describe course of Study:				

EMPLOYMENT

1. Company Name

Telephone
( )–

Address

☐ Full Time
☐ Part Time

Name of Supervisor/Manager

Type of Business:

	Starting	*Final*
Dates:		
Position:		
Salary:		

Reasons for leaving:

Principal Duties:

2. Company Name

Telephone
( )–

Address

☐ Full Time
☐ Part Time

Name of Supervisor/Manager

Type of Business:

	Starting	*Final*
Dates:		
Position:		
Salary:		

Reasons for leaving:

Principal Duties:

3. Company Name

Telephone
( )–

Address

☐ Full Time
☐ Part Time

Name of Supervisor/Manager

Type of Business:

	Starting	*Final*
Dates:		
Position:		
Salary:		

Reasons for leaving:

Principal Duties:

4. Company Name

Telephone
( )–

Address

☐ Full Time
☐ Part Time

Name of Supervisor/Manager

Type of Business:

	Starting	*Final*
Dates:		
Position:		
Salary:		

Reasons for leaving:

Principal Duties:

We may contact the employers listed above unless you indicate otherwise.

DO NOT CONTACT: Employer Number(s): ____________________

FOR EMPLOYERS USE ONLY

REFERENCE CHECK

References Supplied: ☐ Yes ☐ No

Person/Organization Contacted	Results

Tests Administered	Score/Rating	Comments

Interviewer Name	Comments

Explain why you desire this position and why you believe you would do well at it.

I hereby declare that the information provided on this application, or in any attachments provided by me, is true and complete. I understand that if employed, any falsified information or omission of fact on this application may disqualify me from further consideration for employment and may be considered justification for dismissal if discovered at a later date.

I authorize Delmar to investigate the accuracy of any information from any person or organization (unless otherwise noted) and hereby release Delmar and all persons from all claims and liability of any nature arising from such investigation.

Signature of Applicant

Date

Outdoor Recreation Resources

CHAPTER

29

Recreation on Public Lands and Waters

TEST YOUR KNOWLEDGE

Complete the following:

1. As our population grows and as we have more leisure time available, so does the demand for ____________ activities and facilities.
2. City dwellers frequently visit outdoor recreation areas more so than do ____________ people.
3. Huge tracts of land were simply seized from the ____________.
4. The ____________ essentially doubled the size of the nation.
5. The ____________ of 1867 added yet more public domain to the United States.
6. Federal lands relevant to this chapter are managed by which departments?
   a.
   b.
7. List the agencies of the Department of the Interior.
   a.
   b.
   c.
   d.
   e.
8. The primary recreational programs on public lands are operated by:
   a.
   b.
9. Congress set aside Yellowstone National Park in the year ____________.
10. The National ____________ System (NPS) comprises 392 areas.

11. The largest park is Wrangell-St. Elias National Park and Preserve in Alaska, totaling 13.2 million acres.
12. The NPS groups sites into three basic categories:
    a.
    b.
    c.
13. ____________ areas are those areas used to maintain and restore historic wholeness.
14. Historic areas include:
    a.
    b.
    c.
    d.
    e.
    f.
    g.
    h.
15. ____________ areas are used to promote public recreation, and the primary objective is to provide outdoor recreational opportunities.
16. Recreational areas include:
    a.
    b.
    c.
    d.
    e.
17. The park's system contains a number of sites designated as ____________ areas, lands minimally disturbed by humans.
18. In 1964, the federal government started the national ____________ preservation system, which designated 54 national forest areas to remain natural forever.
19. The purpose of the national ____________ system is to provide outdoor recreation by the use of trails.
20. The national recreation trails are usually located near ____________ cities.
21. The national system of wild and scenic rivers classifies the rivers into three categories:
    a.
    b.
    c.
22. The National ____________ System is defined by the Forest Service as "A nationally significant system of federally owned units of forest, range, and related land..."
23. The National Forest System is made up of:
    a.
    b.
    c.
    d.
    e.

24. List the National Forest Service, Special Designated Areas.

 a.

 b.

 c.

 d.

 e.

 f.

 g.

25. The National Forest system is administered by the Forest Service of the U.S. Department of ___________.

26. State recreation areas usually include:

 a.

 b.

 c.

 d.

 e.

 f.

 g.

 h.

 i.

 j.

27. The National ___________ on Recreation and the Environment (NSRE) is a periodic study that examines U.S. outdoor recreation.

28. Moving water and large bodies of water are considered to be part of "The ___________."

29. The ___________ all have their own laws regarding access to the water that we all own.

30. It costs the government ___________ of dollars to repair and clean areas vandalized by inconsiderate people.

ACTIVITY

Purpose:

Evaluate state recreation areas.

Research:

1. Locate state tourism information.
2. Find the areas detailing *state* recreation agencies.

Procedure:

Determine the name for the following state recreation agencies. Detail the information that can be found under the respective headings:

State parks
State recreation areas
State forests
State fish hatcheries
State memorials
State museums
Nature preserves
State historic preservation sites
State fish and wildlife areas

Observations:

1. What agency would you contact for information for damage caused by deer?
2. What agency would you contact for information about camping in a state park?
3. What agency would you contact for information about federal endangered species?

CHAPTER

30

Outdoor Safety

TEST YOUR KNOWLEDGE

Complete the following:

1. As outdoor activities increase and areas become more crowded, the chances for accidents ____________.
2. List three land-based recreational activities from the chapter.

   a.

   b.

   c.
3. Two-thirds of all gun casualties are caused by persons under the age of ____________.
4. Never handle a gun without ____________ supervision.
5. Never let anyone you are with mix ____________, or drugs, and guns.
6. The ten commandments of gun safety include:

   a.

   b.

   c.

   d.

   e.

   f.

   g.

   h.

   i.

   j.
7. Most states require a ____________ to hunt.
8. The money collected from licenses usually is spent on wildlife ____________ measures.

9. A ____________ limit means a hunter can take only a certain number of animals per day or season.
10. Hunter ____________ is a bright, fluorescent color which can be easily seen.
11. ____________ and arrows are used for target practice as well as for hunting.
12. You should never nock an arrow or ____________ a bow if someone is in front of you.
13. The ____________ is the end of the arrow that fits onto the bowstring.
14. Much snowmobiling is done after ____________.
15. All ____________ should adhere to the basic snowmobile safety code.
16. List the basic rules for safe snowmobile operation.
    a.
    b.
    c.
    d.
    e.
    f.
    g.
    h.
    i.
    j.
    k.
    l.
17. ____________ or a change in skin color to gray or yellow-white spots indicates that frostbite has taken place.
18. A ____________ should be aware of basic first-aid procedures.
19. ____________ occurs when the body loses heat faster than it can be produced.
20. A typical first-aid kit contains:
    a.
    b.
    c.
    d.
    e.
21. List the common water-based recreational activities listed in this chapter.
    a.
    b.
    c.
    d.
22. A safe boat contains a ____________ extinguisher.
23. A boat should contain a whistle or ____________ that will blast for at least 2 minutes.
24. A ____________ must contain lifesaving equipment.

25. The four kinds of personal flotation devices listed in the chapter are:
    a.
    b.
    c.
    d.
26. List the common traffic rules on water as listed in the chapter.
    a.
    b.
    c.
    d.
    e.
    f.
27. Common water sports mentioned in the chapter include:
    a.
    b.
    c.
    d.
    e.

ACTIVITY

Purpose:

Determine your fire escape route.

Research:

Walk around your home and make notes on where the windows and doors (internal and external) are located on all levels.

Procedure:

1. Draw a floor plan of your household, including all levels, windows, and doors.
2. Establish one primary escape route.
3. Use arrows to indicate an escape route from each room in the house.

Example:

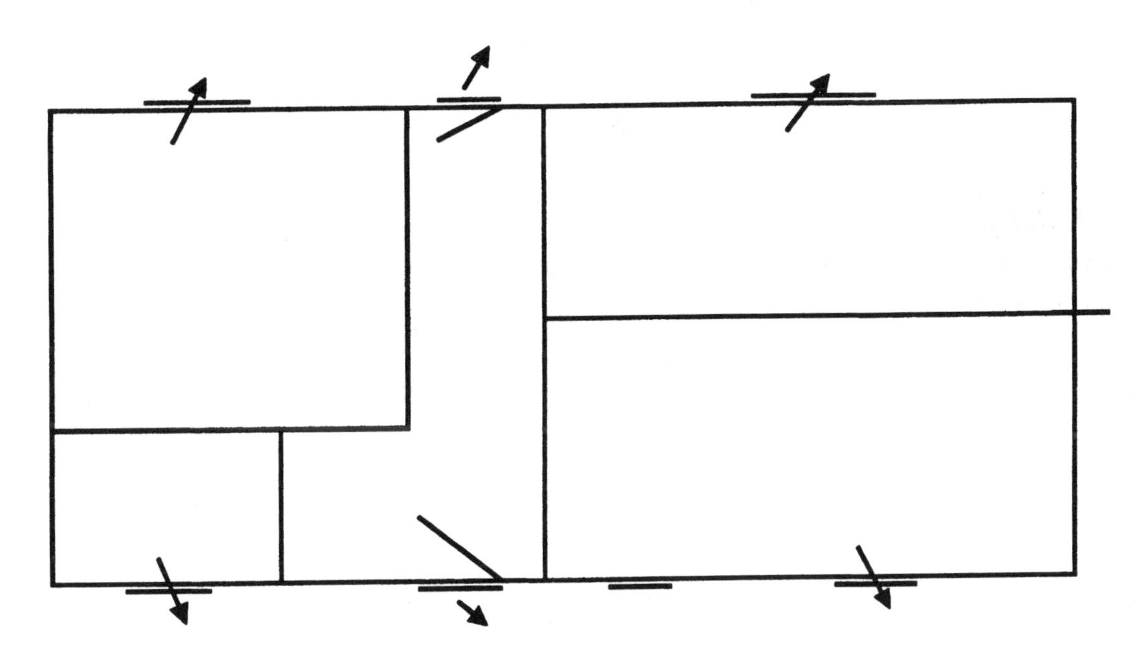

4. Make your plan drawing here:

Observations:

Mark an X in the location outside the house where everyone is to meet in case of a fire. Why is this an important step?

Careers in Outdoor Recreation

TEST YOUR KNOWLEDGE

Complete the following:

1. Careers in outdoor recreation and leisure activities are ____________ in number.
2. Outdoor recreation careers include ____________, professional, skilled, or unskilled jobs.
3. In outdoor recreation, you might be employed by the federal, state, or ____________ government, or private industry, or business.
4. Government employees in outdoor recreation include:

   a.

   b.

   c.

   d.

   e.

   f.

   g.

   h.
5. Park rangers perform professional duties to maintain the parks in ____________ condition.
6. Rangers work to see that areas are free from ____________ hazards and visitor vandalism.
7. Rangers attempt to carry out programs that show the park's visitors the ____________ of the resources preserved in the park.
8. Park rangers may take their knowledge to areas outside the park itself and may be invited as a guest ____________ about the park and its activities at local functions.
9. Park rangers are required to have at least a ____________-year bachelor's degree when entering the career field.
10. As experience increases, rangers may ____________ to district rangers, park managers, or park planning managers.
11. Park rangers must take the federal service examination and qualify under the ____________ Service Commission.

12. Government employees are paid according to a Government ____________, which assigns a grade to the job and considers the experience level of employees and refers to the scale as a general schedule.
13. The National Parks Service (NPS) has many professional, ____________, and trade opportunities across the country in parks and regional offices.
14. The NPS pay scale is based on the ____________ Service pay scale.
15. A jobs ____________ annual salary rating indicates where on the pay scale the worker would fall.
16. List examples of government positions in outdoor recreation or leisure activities, as mentioned in the chapter.
    a.
    b.
    c.
    d.
    e.
    f.
    g.
    h.
    i.
17. To find out whether the NPS might be a place for you, consider doing ____________ work or securing an internship in the summers.
18. The recreational programs in the national wildlife refuges have greatly ____________ in the past years.
19. Recreation managers and workers are employed in planning, coordinating, and developing and implementing ____________ programs.
20. Outdoor recreation managers/workers may be:
    a.
    b.
    c.
    d.
    e.
    f.
    g.
    h.
21. The recreation manager/worker may work in or ____________ a camping area.
22. Recreation managers/workers direct campers in ____________-oriented forms of recreation.
23. Recreation managers/workers are expected to keep ____________ and maintain equipment.
24. List the jobs for a seasonal worker in recreation, as mentioned in the chapter.
    a.
    b.
    c.
    d.
    e.

25. Individuals in recreation careers must be able to ____________ people and be sensitive to their needs.
26. As leisure time increases, so do the ____________ possibilities.
27. According to the U.S. Bureau of Labor Statistics, the median earnings of full-time recreation workers in 2010 was $____________.
28. Workers in full-time positions in the recreation industry generally receive full ____________.
29. Private recreation business opportunities listed in the chapter include:
    a.
    b.
    c.
    d.
    e.
    f.
    g.
    h.
    i.
    j.
    k.
    l.
    m.
    n.
    o.
    p.
    q.
30. It is wise to take a job in your area of interest before ____________ your time and money to start your own business.
31. ____________-time positions give you a chance to explore many options.

ACTIVITY

Purpose:

Evaluate outdoor recreation regulations.

Research:

1. Locate the state game, fish, and parks hunting regulations and/or fishing regulations.
2. Locate the list of game animals.

Procedure:

1. Find the information about each animal in the handbook.
2. Determine the bag limit for each game animal.
3. Determine the season of each game animal.
4. Complete the following chart:

Game animal	Bag limit	Season
1		
2		
3		
4		
5		
6		
7		
8		
9		
10		
11		
12		

Observations:

1. Considering today's date, what is the next hunting season?
2. List four other items of hunting information found in the handbook.
   a.
   b.
   c.
   d.
3. List safety issues in hunting.

JOB EXERCISE FOR CHAPTERS 29–31

Purpose:

Interpret an outdoor recreation activity.

Research:

Choose an outdoor recreation activity and research jobs related to this activity.

Procedure:

Write a detailed report on your chosen outdoor recreational activity and related jobs. Your report should contain the following information:

Introduction

Explain the activity.

What jobs are available?

What training is required?

What equipment is used in the activity?

Where is the activity done?

What is the cost related to performing the activity?

Observations:

1. How are you involved in the activity?
2. How can others become involved in the activity?

UNIT

VII

Energy, Mineral, and Metal Resources

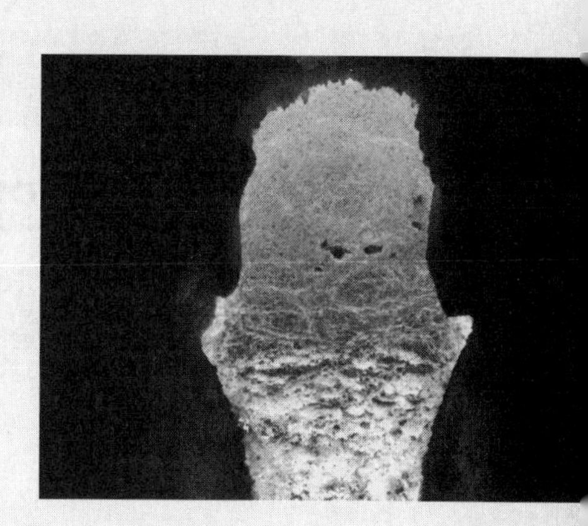

CHAPTER

32

Fossil Fuels

TEST YOUR KNOWLEDGE

Complete the following:

1. The United States developed into a leading nation of the world because of its natural __________.
2. Fossil fuels are minerals formed over time from compressed organic matter, almost exclusively __________.
3. Fossil fuels include:
   a.
   b.
   c.
   d.
   e.
4. Coal is formed from plants in __________ that were covered by deposits of rock or soil.
5. Coal is ranked and classified according to its __________ content.
6. The lowest carbon coal, and the first step of coal formation, is __________.
7. As the rock and soil above the lignite increases, the high pressures cause the coal to become harder; this step in coal formation is called __________ coal.
8. With additional pressure and time, subbituminous coal is transformed into __________ coal.
9. The hardest and oldest coal is __________ coal.
10. Coal was formed about 300 million years ago during a period of time called the __________ period.
11. It takes more than __________ feet of plant matter to make a 1-foot vein of bituminous coal.
12. Most of the coal found today is found in __________ or seams.
13. Coal has been discovered in seams from as small as __________ up to as large as several hundred feet thick.
14. Nearly all the __________ deposits are found between the Appalachian Mountains and the edge of the Great Plains.
15. Coal generates __________ % of all of the world's electricity.

16. As coal burns, a poisonous gas called __________ dioxide is released.
17. Scientists believe that sulfur dioxide contributes to the growing problem of __________ rain.
18. Coal comes from what two types of mines?
    a.
    b.
19. Surface mine coal is taken from a ditch called a __________.
20. The three types of underground mines include:
    a.
    b.
    c.
21. The __________ mine access passageways are vertical to the coal seam.
22. In a __________ mine, the access tunnel is on a slope from the surface to the coal seam.
23. A __________ mine is used when a passageway is bored into a hill or mountain.
24. List the two common methods used to remove coal.
    a.
    b.
25. List the four things that are the cause of most mining accidents.
    a.
    b.
    c.
    d.
26. The U.S. Congress enacted the Federal Coal Mine __________ and Safety Act in 1969 to strengthen the standards for ventilation, coal dust concentration, and mining structure.
27. Coal __________ converts coal into fuel.
28. The United States has the __________ known coal reserves in the world.
29. Petroleum has been called "black __________."
30. A traditional barrel of crude oil contains __________ gallons.
31. Oil requires only __________ million years to form, as compared to coal, which requires several million years.
32. As pressures were exerted on the materials, oil was formed and squeezed into the rock openings or into specialized rocks called __________ rocks.
33. Machines used to find oil include the gravimeter, magnetometer, and the __________.
34. The __________ uses the principle that the gravitational pull of oil-filled rocks differs from rocks containing no oil.
35. The __________ measures differences in the magnetic pull of the earth to find oil-bearing rocks.
36. The __________ uses sound waves to identify various layers and formations under the earth's surface.
37. A hoisting apparatus for the drilling rig is called a __________.

38. Types of offshore operations mentioned in this book include:

    a.

    b.

    c.

    d.

    e.

39. Crude oil is distilled into products including:

    a.

    b.

    c.

40. Fuels include:

    a.

    b.

    c.

    d.

    e.

    f.

    g.

41. Lubricants include:

    a.

    b.

    c.

    d.

42. Petrochemicals include:

    a.

    b.

    c.

    d.

    e.

    f.

    g.

43. As technology has changed over the years, it has become possible to recover crude oil that could not have been __________ earlier.
44. One key estimate of oil for the future might be what is called estimated ultimately __________ (EUR) oil.
45. Over the last 50 years, the world EUR oil reserve has actually __________.
46. Of the 1 trillion barrels that are actually proven to be available, two-thirds are in the __________ Gulf.

47. List the four Persian Gulf countries with the most proven reserves of oil.

    a.

    b.

    c.

    d.

48. Ways to conserve oil listed in this chapter are:

    a.

    b.

    c.

    d.

    e.

49. __________ is a sedimentary rock that contains enough organic material, called kerogen, to yield oil.
50. __________ are a combination of clay, sand, water, and bitumen, a heavy black viscous oil.
51. List the three areas that the gas industry can be divided into.

    a.

    b.

    c.

52. The natural gas is usually located __________ an oil deposit.
53. Natural gas taken from a well must be cleaned in an __________ unit.
54. Natural gas is the end product, but during processing, what three other gases are processed?

    a.

    b.

    c.

55. Natural gas is measured and sold in one of two different units. List those units.

    a.

    b.

56. The common method of measuring natural gas in the United States is __________ feet.
57. When the demand for natural gas is low, the excess can be stored in old __________ wells.
58. Gas that has been cooled to __________ degree(s) F changes to a liquid.
59. Natural gas stored in a liquid state is termed __________ or liquid natural gas.
60. __________ is produced as a by-product of the process of refining petroleum.
61. __________ involves injecting water, chemicals, and either sand, glass beads, or aluminum balls under high pressure through a drilled well to break up rocks underlying the surface.
62. In May 2013, the Energy Information Administration (EIA) predicted that the United States could become energy __________ by 2030.

ACTIVITY

Purpose:

Calculate miles per gallon.

Research:

1. The formula for miles per gallon is:

$$\frac{\text{Miles}}{\text{Gallons}} = \text{Miles per gallon (round to the nearest 10th)}$$

2. The formula for miles is:

$$\text{Gallons} \times \text{Miles per gallon} = \text{Miles (round to the nearest mile)}$$

3. The formula for gallons is:

$$\frac{\text{Miles}}{\text{Miles per gallon}} = \text{Gallons (round to nearest 100th)}$$

Procedure:

1. Using the formula provided and the information included in the first Table below, calculate miles per gallon.
2. Using the formula provided and the information included in the second Table below, calculate miles.
3. Using the formula provided and the information included in the last Table below, calculate gallons.

Calculate miles per gallon

Trip	Miles	Gallons	Miles/gallon
1	100	10	
2	200	12	
3	350	20	
4	1,250	58	
5	1,500	100	
6	3,000	120	

Calculate miles

Trip	Miles	Gallons	Miles/gallon
1		9.7	18
2		11.2	21
3		27.5	22
4		20.6	25
5		106.5	27
6		142.2	32

Calculate gallons

Trip	Miles	Gallons	Miles/gallon
1	250		12
2	1,500		15
3	750		17
4	600		21
5	500		20
6	1,350		38

Observations:

Show your calculations on a separate sheet.

Alternative Energy Sources Management

TEST YOUR KNOWLEDGE

Complete the following:

1. As the ____________ of fossil fuels increase, many Americans are searching for other ways to supply their energy needs.
2. List the main types of alternative energy sources in use today, as mentioned in the chapter.
   a.
   b.
   c.
   d.
   e.
   f.
   g.
   h.
   i.
3. Solar energy is more abundant, less exhaustible, and more ____________ -free than any other energy source.
4. Solar energy systems can be of the ____________ and passive type.
5. Active systems ____________, store, and distribute the energy from the sun.
6. Passive systems rely on the natural airflow for ____________.
7. Parts of an active system include:
   a.
   b.
   c.

8. The active system collects the sun's energy in the form of ____________ and stores it for future use.
9. A passive solar system has only a ____________ device.
10. The first ____________ cell was developed in the 1880s.
11. Research is being done to convert sunlight to electricity by the use of solar ____________.
12. ____________ power plants have had much news coverage in recent years, mostly because of fears of a nuclear disaster.
13. All matter is composed of small, submicroscopic particles called ____________.
14. Radium emits:

    a.

    b.

    c.
15. When rays interact with other compounds, they split atomic nuclei, in a process called ____________.
16. During the fission process, ____________ is given off, which is the important component in a nuclear power plant.
17. Fissioning one pound of uranium slowly can produce ____________ million kilowatt-hours of power.
18. The heart of the nuclear power plant is the ____________.
19. The reactor uses a mixture of U235 and U238, in the form of ____________ pellets, as a fuel source.
20. To control or stop the reaction, the reactor uses ____________ rods to absorb neutrons.
21. Water in tubes surrounding the nuclear core turns to ____________, which is sent through turbines to turn electrical generators.
22. The main concern with the increased use of nuclear power is the fear of an explosion or -uncontrolled heat buildup causing a ____________.
23. A concern with nuclear power plants is the disposal of ____________ wastes from the reactor.
24. Nuclear wastes are packaged in ____________ -steel containers and buried.
25. ____________ power generation involves capturing heat from the earth to produce electricity.
26. The two main disadvantages of geothermal energy are:

    a.

    b.
27. ____________ is the total dry weight of all the living organisms in a given area at a given time.
28. Plants capture ____________ from sunlight during photosynthesis.
29. List the two types of biomass.

    a.

    b.
30. A gasoline/ethanol blend used to power your automobile is referred to as "____________."
31. Alcohol is ____________ explosive and more stable than gasoline.
32. Alcohol is produced by growing yeasts in a ____________ solution.
33. Yeasts take in sugar, proteins, vitamins, and minerals and give off carbon dioxide and ____________.
34. The most commonly used grain for ethanol production is ____________.
35. The residue from the alcohol can be used as ____________ for livestock.
36. ____________ bacteria grow and prosper in the absence of oxygen.
37. Methane is sometimes referred to as "____________."
38. Methane is a by-product of decomposition by ____________ bacteria.

39. ____________ is an odorless gas with a heating rate of 600–700 British thermal units (BTUs) per cubic foot of gas.
40. 1 BTU is approximately the amount of heat needed to raise 1 pound of ____________ 1 degree F.
41. Decomposing wastes produce methane and a compound called hydrogen ____________.
42. Hydrogen sulfide gives the gas a very unpleasant ____________.
43. Methane gas can be produced artificially in a device called a methane ____________.
44. After the digester wastes have been digested, solids called ____________ remain.
45. Sludge is an important by-product of methane production because of its use as a ____________.
46. Methane cannot be highly ____________ like other gases.
47. Sugarcane and sugar beets are easy to convert to energy through ____________.
48. ____________ are a potential energy crop.
49. ____________ crops have good potential for future biomass energy production.
50. Soybean and sunflowers produce oil that can be converted to ____________.
51. ____________ that grow well in lakes and ponds can be used to produce oil that can then be converted to fuel.
52. The economic effect of rising petroleum prices has made ____________ ethanol production a high priority in the United States Department of Agriculture (USDA) research.
53. ____________ is particularly interesting for energy production because it thrives in areas where regular agricultural crops do not work well.
54. ____________ is waterpower.
55. The power from moving water is used to do ____________.
56. The potential of ____________ power is estimated at 2.9 million megawatts of power.
57. Wind power requires the use of a ____________ device like a DC battery.
58. Wood is a good source of heat energy, but it is less ____________ to burn.
59. The Federal Energy Policy Act of 2005 includes what two important provisions?

    a.

    b.

ACTIVITY

Purpose:

Calculate the electrical costs for appliances.

Research:

1. A watt-hour is 1 watt for 1 hour.
2. A kilowatt-hour indicates the use of 1,000 watts of electricity for 1 hour.
3. The cost of residential electricity is charged by the kilowatt-hour.

$$\frac{\text{Wattage} \times \text{time in hours}}{100} = \text{Price per kilowatt} = \text{Electricity cost}$$

Example:

$$\frac{120 \times 2}{100} = 24$$

$$0.24 \times 0.01 = 0.024$$

Procedure:

Complete the following table to determine the cost:

Appliance	Watts	Time	Price	Cost
Range	12,000	1 hour	.10	
Dishwasher	1,200	3 hour	.10	
Coffeemaker	1,100	8 hour	.10	
Washer	1,200	5 hour	.10	
Dryer	9,000	5 hour	.10	
Water heater	5,000	10 hour	.10	

CHAPTER

34

Metals and Minerals

TEST YOUR KNOWLEDGE

Complete the following:

1. Our mineral and metal resources are __________ resources.
2. The earth contains about __________ billion tons of metal resources per square mile of land surface.
3. Problems stem from the fact that the mineral resources are at such __________ concentrations and at such depths that it is not usually economically feasible to remove them.
4. Ferrous means __________ containing.
5. Of the __________ metals, iron ore is by far the most important.
6. The __________ industry is the foundation of all modern industry.
7. Discoveries of large amounts of iron ore in the __________ region spurred tremendous growth of the steel industry in Indiana, Illinois, New York, and Minnesota at one time.
8. The kinds and varieties of iron ore listed in the chapter are:

   a.

   b.

   c.

   d.
9. Low-grade ores are called __________.
10. A high-quality ore contains at least __________ % metallic iron.
11. When iron ore is processed, __________ is made.
12. The __________ process lead to a cheap, practical production of steel.
13. Separating the ore from the rock can be called:

    a.

    b.

    c.

    d.

14. Smelting is melting the iron ore to remove __________ from the concentrate.
15. Once the molten iron is processed, it is poured into molds to form bars called __________ iron.
16. Pig iron bars are processed into what forms of steel?
    a.
    b.
    c.
    d.
17. When other ferrous metals are combined with iron, they are called __________.
18. Ferroalloys include:
    a.
    b.
    c.
    d.
    e.
    f.
19. Nonferrous metals include:
    a.
    b.
    c.
    d.
    e.
    f.
20. __________ is commonly used for tools and weapons.
21. Copper is used in kitchen utensils, __________, screens, and piping.
22. One-half of all copper is used in __________ products.
23. Copper and tin make __________.
24. When copper is combined with __________, we call it brass.
25. Much of the copper used is __________.
26. __________ is one of the most common metals known on the earth.
27. List the reasons aluminum is often used in place of steel.
    a.
    b.
    c.
    d.
    e.
28. Although other ores such as kaolin and corundum contain aluminum, __________ is the easiest to find and mine.
29. Bauxite ore contains a large amount of aluminum __________.
30. Once the bauxite ore reaches the processing plant, the aluminum oxide is extracted from the bauxite in a process called the __________ process.

31. Nearly all bauxite used in the United States is __________.
32. Because of the large amount of electricity needed to produce aluminum from bauxite, __________ aluminum is important.
33. __________ is used in batteries and metal building construction.
34. About 60% of the lead used comes from __________ sources.
35. The most common use of zinc is for galvanizing steel, thus making it rust __________.
36. Tin is combined with copper to make __________.
37. __________ is the only metal stable in a liquid state at ordinary temperature.
38. The main uses of mercury are for __________ equipment, special paints, and industrial chemicals.
39. Plant minerals are called __________.
40. Plant minerals are an important part of producing food and include:

    a.

    b.

    c.

41. Most nitrogen used is produced by a __________ process from air.
42. Phosphorus is found in nature in the form of __________ rock.
43. Potassium occurs in the form of __________ in beds similar to coal.

ACTIVITY

Purpose:

Perform a hardness test.

Procedure:

1. Locate a hardness test kit.
2. Locate the samples provided by the instructor.

Observations:

1. Follow the instructions in the kit to make hardness tests on the samples.
2. Complete the following table:

Sample	Hardness
1	
2	
3	
4	
5	

CHAPTER

35

Careers in Energy, Mineral, and Metal Resources

TEST YOUR KNOWLEDGE

Complete the following:

1. ____________ is one of the major concerns in this century.
2. According to the U.S. Department of Labor (2005), oil and natural gas provide us with ____________ of our energy needs.
3. With the public pressure to lower U.S. dependence on foreign oil, many ____________ have opened up in the energy career areas.
4. List the five areas in which a person employed in the oil and natural gas field can work, as discussed in this chapter.
   a.
   b.
   c.
   d.
   e.
5. The ____________ process is usually accomplished by a small crew of workers who search for and study geological formations that might contain oil.
6. The exploration team is led by a petroleum ____________ who analyzes and interprets the information gathered by the team.
7. ____________ study fossil remains to locate oil-bearing layer rock.
8. ____________ study the physical and chemical properties of mineral and rock samples.
9. ____________ determine the rock layers most likely to contain oil and natural gas.
10. ____________ examine and interpret aerial photographs of land surfaces.
11. A ____________ investigates the history of the formation of the earth's crust.

12. Specialists on the exploration team are called ____________ and surveyors assisting in surveying and mapping operations.
13. A geophysical prospector usually leads a crew who operates and maintains electronic ____________ equipment.
14. Seismic equipment is used to investigate and collect ____________ leading to the possible discovery of new gas or petroleum deposits.
15. Once oil is found or a likely area is located, the ____________ team is assembled.
16. The overall planning of the drilling operation is the responsibility of a petroleum ____________.
17. The petroleum engineer oversees the drilling process and offers ____________ advice.
18. A typical drilling crew includes:

    a.

    b.

    c.

    d.
19. The ____________ supervises the crew, operates machinery that controls drilling speed and pressure, and records operations.
20. The ____________ operator is responsible for the fine operation and general maintenance of the mud, pumps, and machines used in the pump room.
21. The helpers, commonly referred to as ____________, generally work on the rig floor.
22. Associated with the drilling team are general laborers, commonly referred to as ____________, who do general oil-field maintenance and construction work, such as cleaning tanks and building roads.
23. ____________ open and close valves to regulate the oil flow from wells to tanks or pipelines.
24. ____________ measure and record the oil flow and take samples to check quality.
25. ____________ test the oil for water and sediment and remove these impurities when they are found.
26. Skilled workers involved in the petroleum industry include:

    a.

    b.

    c.

    d.
27. The gas ____________ operates a unit that removes unwanted components from the natural gas.
28. The ____________ station operator tends compressors that raise the pressure of the gas for transmission through the pipeline.
29. The oil well ____________ mixes and pumps cement into the space between the steel casing and the well walls, thus preventing cave-ins.
30. An ____________ pumps acid down the well to increase the flow of oil.
31. The ____________ operator uses a subsurface gun to pierce holes in the casing to make openings so that oil flows into the well bore.
32. The ____________ taker operator takes samples for the geologist to test.
33. The well ____________ clean, repair, and salvage the well parts.
34. Most jobs in the petroleum field are rugged, ____________ jobs.
35. Workers in the oil and gas exploration crew are usually trained in the ____________.
36. Scientists such as geologists and geophysicists need ____________ training with at least a bachelor's degree.
37. Employment in petroleum and natural gas production in the United States is expected to ____________.

38. The petroleum engineer occupation is expected to experience a 25% employment ____________ from 2012 to 2022.
39. About 85% of the people in the coal industry are ____________ workers.
40. List the two methods of obtaining underground coal.
    a.
    b.
41. In ____________ mining, the cutting-machine operator uses a huge electric chain saw to cut a strip underneath the coal seam.
42. The ____________ mining method eliminates the drilling and blasting operations of conventional mining.
43. Before miners are allowed underground, a mine ____________ inspector checks the work area for loose roofs, dangerous gases, inadequate ventilation, and other hazards.
44. The ____________ -dust machine operator sprays limestone on the mine walls and ground to hold down the explosive, breath-impairing coal dust.
45. The roof ____________ operates a machine that installs roof-support bolts to prevent cave-ins.
46. The ____________ builder constructs doors, walls, or partitions in the passageways to force air through the tunnels and into the work areas.
47. When the earth and stone are removed above the coal seam, it is called ____________ mining.
48. In many strip mines, the ____________ is first drilled and blasted.
49. Once the overburden is removed, the coal-loading ____________ operator rips the coal from the seam and loads the coal into trucks to be driven to the preparation plant.
50. At the ____________ plant, rocks and other impurities must be removed before the coal is crushed, sized, or blended to meet consumer needs.
51. Environmental scientists and ____________ identify locations that will yield enough coal to justify the cost of extracting the coal.
52. Mining ____________ examine the coal seams.
53. ____________ engineers oversee the installation of equipment.
54. ____________ engineers are in charge of health and safety programs.
55. Technical personnel in the coal industry include:
    a.
    b.
56. Miners are subject to ____________ conditions.
57. Most miners start out as helpers and advance to more skilled jobs with ____________.
58. List the careers of the nuclear energy industry, as mentioned in the chapter.
    a.
    b.
    c.
    d.
    e.
    f.
59. Nuclear ____________ conduct research and design the processors, systems, and instruments to use nuclear energy.
60. Nuclear equipment monitoring ____________ monitor the results of nuclear experiments and the effects on humans, facilities, and the environment.

61. Nuclear equipment ____________ technicians assist nuclear scientists in laboratory and production activities by operating the equipment needed to release, control, and use nuclear energy.

62. Power plant ____________ work in the power-generating plants to control and monitor boilers, turbines, generators, and auxiliary equipment.

63. Power ____________ and dispatchers control the flow of electricity through transmission lines to locations that supply the needs for electricity.

64. Except for ____________ safety precautions, the working conditions of the nuclear industry are the same as in other industries.

65. Specialized ____________ of nuclear energy is essential for most workers.

66. There is expected to be ____________ change in nuclear power plant operators, distributors, and dispatchers from 2010 to 2020.

67. The steel industry's employment is broken into two major sectors: iron and steel mills and ____________ production.

68. Steel work varies with the type of furnace used and includes everything from unloading raw iron ore to ____________ the steel for shipment.

69. Steel work is hazardous; with an injury and illness rate almost ____________ as high as for all other industries.

ACTIVITY

Purpose:

Answer interview questions.

Research:

1. Review the following information about what is needed at, and what to know about, interviews:
   Bring a pen
   Bring personal information including your résumé
   Bring your social security number
   Be prepared to discuss your specific skills
   Be positive
   Never answer with just "yes" or "no"
   Answer all questions to the best of your ability
   Do not be afraid to say you do not know
   Answer all questions honestly
   Be prepared to ask two good questions at the end of the interview
   Discuss salary range at the very end of the interview
2. Choose a job in the energy, metal, and mineral industry and research the job requirements.

Procedure:

Answer the following questions as if you were applying for the job you chose above:

1. What are your future plans?
2. To what organizations do you belong?
3. Why do you think you would like to work for this company?
4. Do you have any special abilities or skills?
5. Do you have any similar work experience?
6. Do you have any hobbies?
7. What do you want to be doing 5 years from now?
8. Why do you want this job?
9. Why do you think you are qualified for this job?
10. What are your outside interests?
11. What have you done that shows initiative and willingness to work?

Observations:

What are the two questions you would ask an interviewer at the end of an interview?

JOB EXERCISE FOR CHAPTERS 32–35

Purpose:

Write a letter of application.

Research:

When applying for any position you should write a letter of application (see Example). The first paragraph should

a. Tell the employer exactly what job you are applying for.
b. Tell where you heard about the position.
c. Include an expression of interest in the company.

The second paragraph should

a. Sell yourself by highlighting your main accomplishments.
b. Provide your current employment information. This should be kept brief because this information is in your résumé.
c. List important educational information.

The third paragraph should

a. Note whether you have enclosed supplemental information.
b. Include your telephone number and when it is best to contact you.
c. Re-express your interest in the job.
d. Thank the employer for their time.

Procedure:

Write a letter of application for one of the following jobs:

petroleum engineer
steel industry salesperson
dragline operator
miner
engineering technician
drilling crew laborer
steel industry laborer
gas treater
soil conservation technician
oceanographer
geologist
forestry technician
game biologist
fishery biologist
fish and wildlife technician

Example:

November 18, 20__

433 South 108th Street
Rapid City, SD 57701

Gerri Ann Erdeen
Job Placement Officer
Hometown, SD 57000

Dear Ms. Erdeen:

I am interested in the position of Assistant Tree Trimmer that you have advertised in the *Rapid City Journal.* Many of your customers have told me that you provide prompt service and a positive customer atmosphere. I would like to become a part of such a company.

Through the years, I have improved my ability in the landscape field. In my agriculture class I studied tree identification, proper procedures to prune trees, and many other job-related skills that have proven important in work situations. I have also done projects in which I was responsible for the overall maintenance of the lawn and garden. Through these projects I have learned to properly mow the lawn; prune plant materials; keep the garden in shape by deadheading and replanting, and spraying chemicals; and provide customer relations. I hope to use these skills to become a responsible employee that is able to do the main duties of this position.

This past year I competed in the Nursery Landscape contest and placed as the top individual in our state. The team also placed first. This contest consisted of a general knowledge test, reading a landscape plant, and identifying fifty trees using only the bark and the buds. Recently, I competed at the National FFA Convention in Indianapolis, Indiana against students from across the United States and I received a silver ranking. I am also currently the Rapid City FFA Chapter Treasurer. I believe that this leadership role has shown that I have the determination and leadership that will help me become successful in your company.

Please find my enclosed resume containing further information. I am excited about the possibility of working for your company and look forward to hearing from you at your convenience. You may contact me at 605-351-5555 or via my email astudent@mail.com.

Thank you for your time and consideration.

Sincerely,

Andy Student

UNIT

VIII

Advanced Concepts

CHAPTER

36

Advanced Concepts in Natural Resources Management

TEST YOUR KNOWLEDGE

Complete the following:

1. ____________ implies doing something.
2. ____________ an area in its natural state is an intentional management practice.
3. The first step in management is ____________ making.
4. Selecting the ____________ management practice from among alternative solutions is a complex process.
5. Making a decision about the best management practice consumes a great deal of ____________ and effort on the part of local, state, and national agencies.
6. Our natural resources are held in ____________ by the government for the benefit of the people.
7. An important change in natural resources management has been that the public is allowed, even encouraged, to ____________ in the decision-making process.
8. A great deal of controversy surrounds whether the public is capable of making valid, well-informed ____________ regarding management of the public's natural resources.
9. Public participation is ____________ mandated.
10. A continuum, or scale, labeled "exploitation" on one end and "preservation" on the other represents the spectrum of conservation ____________.
11. List the viewpoints of conservation philosophies, as described in the chapter.
    a.
    b.
    c.
    d.

12. ____________ advocate the management of nature to maintain it in a good state while using the resources to benefit people.
13. ____________ change some aspect of nature to make it more valuable and profitable.
14. ____________ take something directly from nature and move it for a profit.
15. Management goals ____________ over time.
16. People have ____________ needs and wants.
17. Natural resources are scarce—they are ____________.
18. The ____________ we make as a society must consider our wants and needs while considering that our resources need protection, maintenance, and enhancement.
19. The ____________ paradox—There Is no Such Thing as a Free Lunch.
20. To gain something in terms of natural resources management, we also ____________ up something.
21. The value of what we give up when we choose one alternative over another is called ____________ cost.
22. ____________ means human centered.
23. Resource ____________ is best assigned according to the use of the resources.
24. Water, soil, and air are considered ____________ resources.
25. The goal of a ____________ yield management is to protect the quantity and quality of the resource.
26. Management for a sustained yield means the use of renewable resources in such a way as to allow a constant rate of ____________ indefinitely.
27. The three Es of resources management are:
    a.
    b.
    c.
28. List the two primary industries in broad terms, as listed in this book.
    a.
    b.
29. As defined in this chapter, agriculture includes:
    a.
    b.
    c.
    d.
30. Mining includes all forms of resource exploitation such as:
    a.
    b.
    c.
    d.
31. Resources are not equally ____________ around the world.
32. Common ____________ are in theory owned by everyone but in reality owned by no one.
33. Whether the resources are owned privately or not has major implications for ____________.
34. Everything we do with natural resources is done with the interests of ____________ at heart.

35. ____________ models are complex systems for visualizing our resource base, especially the complex and dynamic interactions occurring between and among the variables and components.

36. ____________ population is the number one environmental problem.

37. We live in a(n) ____________ ecosystem—our planet.

38. There are no "best" or "right" answers to natural resources management questions, but there are lots of "wrong" ____________.

ACTIVITY

Purpose:

Determine direction using a compass.

Materials Needed:

Compass

100-foot tape

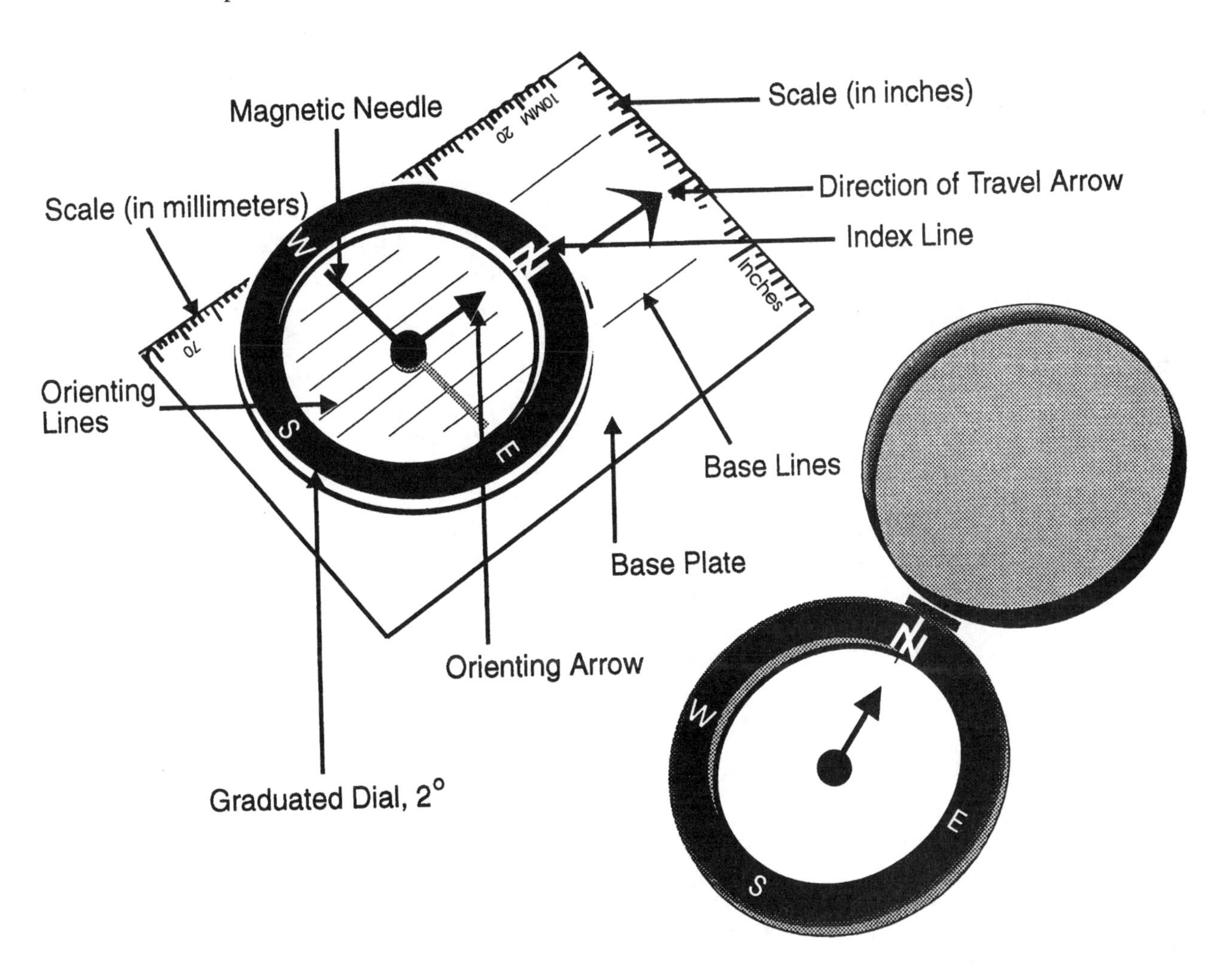

Procedure:

1. Locate the site markers and starting point prepared by the instructor.
2. Stand at Site 1.
3. Hold the compass in your hand in the direction of Site 2.
4. Rotate the compass dial until the orienting arrow overlaps the floating magnetic needle.
5. Read the bearing in degrees from the dial.
6. Measure the distance with the tape.
7. Move to the next point.
8. Repeat the procedure for all site markers.

Site	Bearing	Distance
1		
2		
3		
4		
5		

Observations:

1. In what applications might you use a compass?
2. What groups of people might use a compass?

JOB EXERCISE FOR CHAPTER 36

Purpose:

Complete a résumé.

Research:

A résumé is an advertisement of facts about yourself and your skills. It should contain accurate facts. Your résumé should be on 8- by 11-inch paper and no more than two pages. You should include personal information, educational information, work-related information, and references. Your name should appear prominently at the top of the résumé.

Procedure:

Obtain résumés from people you know. Study the various forms. Choose a format that best suits you. One example is listed on the following page.

Example:

Andy Student
1234 Pandora Circle
Rapid City, SD 57701

605-555-1212
astudent@email.com

Objective: Seeking a position as an assistant greenskeeper in the Rapid City area.

Work History: Presently employed as a gardener and lawn care maintenance person for customers in the area. Responsible for overall maintenance of the gardens and surrounding yard.

Employed at Walters Discount Center, Rapid City, SD. Duties require me to handle customer complaints, and to know cash register operations and the store policies along with lawn and garden center activities. (August 20__ to Present)

Landscaper of an outdoor living area for Dakota Hills Properties, Rapid City, SD. Skills included: laying sod, planting different plant materials, building a rock waterfall, installing rock edging, and pruning trees and shrubs. (September 20__ to October 20__)

Education: Central High School, 433 North Street, Rapid City, SD 57701 (September 20__ to Present)

Academic preparation consisted of courses such as: Computer Aided Drafting, Graphic Communications, and advanced Mathematics classes. Agriculture education classes included Forestry/Natural Resources, Conservation/Natural Resources, Wildlife/Natural Resources, and Agricultural Sales and Marketing.

Activities: FFA, National Honors Society, Toastmasters, District FFA Treasurer, Chapter FFA President; Honors include: 1st place state Nursery/Landscape team and top individual and 1st place Floriculture team.

References:

Mike Laiter
Retail Sales
1222 3rd Street
Rapid City, SD 57702
605-555-2222

Robert S. Wailon
Counselor
433 North Street
Rapid City, SD 57702
605-555-3333

Please Note: Next to References you may also say "Available upon request." If you do choose to list your references, make sure that you have the person's permission before you list his or her information on your résumé.

Observation:

Write your résumé.